Obsessive
Genius

GENERAL EDITORS: EDWIN BARBER AND JESSE COHEN

BY BARBARA GOLDSMITH

The Straw Man

Little Gloria . . . Happy At Last

Johnson v. Johnson

*Other Powers: The Age of Suffrage, Spiritualism, and the
Scandalous Victoria Woodhull*

Obsessive Genius: The Inner World of Marie Curie

GREAT DISCOVERIES

BARBARA GOLDSMITH

Obsessive Genius

The Inner World of Marie Curie

ATLAS BOOKS

W. W. NORTON & COMPANY

NEW YORK · LONDON

Acknowledgment is made to quote material from *Madame Curie* by Eve Curie, translated by Vincent Sheean, copyright 1937 by Doubleday, a division of Random House, Inc. Used by permission of Doubleday, a division of Random House, Inc.

For information about permission to reproduce selections from this book, write to Permissions, W. W. Norton & Company, Inc., 500 Fifth Avenue, New York, NY 10110

Manufacturing by RR Donnelley, Harrisonburg Division
Book design by Chris Welch
Production manager: Julia Druskin

Library of Congress Cataloging-in-Publication Data

Goldsmith, Barbara.
Obsessive genius : the inner world of Marie Curie / Barbara Goldsmith.
    p. cm. — (Great discoveries)
Includes bibliographical references.
ISBN 0-393-05137-4 (hardcover)
1. Curie, Marie, 1867–1934. 2. Women chemists—Poland—Biography.
I. Title. II. Series.
QD22.C8G56 2004
540'.92—dc22

2004015027

W. W. Norton & Company, Inc., 500 Fifth Avenue, New York, N.Y. 10110
www.wwnorton.com

W. W. Norton & Company Ltd., Castle House, 75/76 Wells Street, London W1T 3QT

3 4 5 6 7 8 9 0

For these children

Evelyn
Gillian
Jack
Jenna
Lillian
Max
and their "spirit of adventure"

Contents

Obsessive
Genius

Introduction

Paris, April 20, 1995: the white carpet stretched block after block down the rue Soufflot ending in front of the Panthéon, which was draped in tricolor banners that extended from the dome to the pavement. To the tune of the "Marseillaise," the Republican Guardsmen moved along the white expanse. The thousands who lined the streets were unusually quiet; some tossed flowers as the cortège moved by. Faculty from the Curie Institute were followed by Parisian high school students who held aloft four-foot blue, white, and red letters—the Greek symbols for alpha, beta, and gamma rays.

As the marchers approached the Panthéon, they fanned out and stared up at a platform under the grand dome on which were seated such luminaries as President François Mitterrand. Suffering from cancer, in the last weeks of his fourteen-year presidency, he had decided to dedicate his final speech to the "women of France" and in a dramatic gesture to place the ashes of Madame Curie and her husband, Pierre, in the Panthéon, thus making Marie (Marya Salomee Sklodowska) Curie the first woman to be buried there for her own accomplishments. The Curies had been exhumed from their graves in the suburb of Sceaux to join such French immortals as Honoré-Gabriel Riqueti (Comte de Mirabeau), Jean-Jacques Rousseau, Émile Zola, Victor Hugo, Voltaire (François-Marie Arouet), Jean-Baptiste Perrin, and Paul Langevin.

Seated beside Mitterrand was Lech Walesa, the president of Poland, Madame Curie's native country. Finally, there were the families of the two scientists being enshrined: their daughter Eve, and the children of their deceased daughter Irène and her husband Frédéric Joliot-Curie—Hélène Langevin-Joliot and Pierre Joliot, both distinguished scientists.

Pierre-Gilles de Gennes—the director of the School of Industrial Physics and Chemistry of the City of Paris (EPCI), where Marie and Pierre had discovered radioactivity, radium, and polonium—spoke first, saying that the Curies represented "the collective memory of the people of France and the beauty of self-sacrifice." Lech Walesa spoke of Marie Curie's Polish origins and termed her a patriot both of Poland and France. François Mitterrand then rose:

> This transfer of Pierre and Marie Curie's ashes to our most sacred sanctuary is not only an act of remembrance, but also an act in which France affirms her faith in science, in research, and we affirm our respect for those whom we consecrate here, their force and their lives. Today's ceremony is a deliberate outreach on our part from the Panthéon to the first lady of our honored history. It is another symbol that captures the attention of our nation and the

exemplary struggle of a woman who decided to impose her abilities in a society where abilities, intellectual exploration, and public responsibility were reserved for men.

Above Mitterrand's head, as he spoke these words, one could read the inscription on the Panthéon's façade: TO GREAT MEN FROM A GRATEFUL COUNTRY. The irony is apparent.

After the speeches, a deafening ovation echoed down the streets. The modest Pierre Curie, who wanted to be buried in Sceaux because he abhorred "noise and ceremonies," would have hated this display. But, like it or not, the Curies, especially Marie, had been deified. Madame Curie was now an icon for the ages and an inspiration to women who saw in her the fulfillment of their own dreams and aspirations, however vague. I was among them.

As a teenager, I tacked up, among the detritus on my bulletin board, between a reproduction of Van Gogh's *Starry Night* and my Friday night bowling card, a photograph of Marie Curie sitting under an elm tree, her arms stretched out to encircle the waists of her daughters, two year old Eve, and nine year old Irène. I don't know why I was drawn to this photograph, but it wasn't about science. Madame Curie was my idol, and like other idols you don't have to know what they did to worship them. Perhaps I found comfort in what I took to be Marie's protective embrace since, at the time, my own mother was far away in a hospital, having been critically injured in an automobile accident. Who knows?

In the photograph there are none of the usual smiling faces. All three look ineffably sad. I didn't know why then. Now I do. Under the picture I had placed two of Madame Curie's quotations: "Nothing in life is to be feared. It is only to be

Marie with Eve (left) and Irène (right) in the garden at Sceaux in 1908.

understood," and "It is important to make a dream of life and of a dream reality." It was only while researching this book that I found that Pierre Curie, not Marie, wrote the latter.

In any case, there is little doubt that Marie Curie's life was truly inspirational. She was rare as a unicorn in the field of science. She came from an impoverished Polish family and worked for eight years to earn enough money to study at the Sorbonne. She overcame incredible hardships. In 1893, Marie Curie became the first woman to secure a degree in physics at the Sorbonne. The following year she received a second degree, in mathematics. She was the first woman to be appointed a professor at the Sorbonne, and the first woman to receive not only one but two Nobel Prizes: the first in physics along with her husband, Pierre, and Henri Becquerel (for the discovery of radioactivity); the second, eight years later, in chemistry (for the isolation of the elements polonium and radium). She was the first woman to be elected to the 224-year-old French Academy of Medicine. In addition to having a spectacular career, Marie raised two daughters largely as a single mother and saw that they were well educated, physically strong, and independent.

These are the facts of Madame Curie's life. But, they have become shrouded in a romantic myth constructed to suit the beliefs and proclivities of many people—journalists, scientists, medical practitioners, feminists, businessmen, industrialists, and even Madame Curie herself. She is remembered as a scientific Joan of Arc. Paris streets are named after Madame Curie and her husband, Pierre; the French 500-franc note (now a collector's item) is imprinted with her face and her so-called "miserable shed" laboratory, as well as several scenes from her life. Stamps and coins bear her image. The World War I automobiles that were refitted to contain X-ray equipment were known as "Les Petites Curie." Semi-documentaries and feature films contribute to her legend. As a child, I was entranced with Greer Garson as Marie and Walter Pidgeon as her husband, Pierre, in the 1943 film *Madame Curie*. I remember the movie-star Marie, her face glistening with sweat as she stirs a vat of boiling ore. I will never forget the scene in the dark of the night, when Marie and Pierre enter her laboratory to see a tiny luminous stain congealed in a dish. "Oh Pierre! Could it be? Could this be it?" exclaims Marie, as tears roll down her cheeks. Yes, this was it—Radium!

Many years have passed since a foolish girl was inspired by a Hollywood heroine. Now it is women, and the mores and history of the time in which they lived, that constitute a major theme in my writing. Why are some women trapped in their environment while others escape, or circumvent, or ignore these obstacles? How did society and family affect their aspirations? Why do some women seek independence while others want to tread a prescribed path? And, what common chord did Madame Curie strike in the psyche, particularly of women? These were a few of the questions that intrigued me.

My own obsession lies in an investigation of the vast disparity between image and reality. The mythical Madame Curie remains perhaps the most famous woman scientist in the world. Radium is thought to be her colossal discovery and has been given an enormous importance in the cure of cancers through radiation therapy. But in reality, is this true, and was this her major contribution to science? There is no doubt that over the last century Madame Curie's life has evolved into an image of towering perfection. But behind this image there was a real woman. It was this person that I wished to pursue.

Publicity image of Greer Garson and Walter Pidgeon from the feature film Madame Curie, *1943.*

CHAPTER 1

Early Influences

"A great discovery does not issue from a scientist's brain ready-made, like Minerva springing fully armed from Jupiter's head; it is the fruit of an accumulation of preliminary work," wrote Marie Curie. And Louis Pasteur famously said, "Fortune favors the prepared mind." But major achievements need more than scientific preparedness; they need an individual peculiarly suited to the task. Marie Curie's character, formed by discrimination and deprivation, by parental pressure and ambition, by patriotism and dissembling, was such an individual.

As a child of four she stood transfixed before a glass cabinet: Inside were "several shelves laden with surprising and graceful instruments: glass tubes, small scales, specimens of minerals, and even a gold-leaf electroscope." Professor Wladyslaw Sklodowski told his daughter that the cabinet contained his "physics apparatus." Marya Salomee Sklodowska, nicknamed Manya, who was to become the world-famous Madame Curie, had no idea what these words meant, but "she did not forget." This anecdote, written by Eve Curie, suggests an early attachment to science, but in fact it tells us more about the child's attachment to her father. The glass case remained locked because Professor Sklodowski's science classes had been canceled after the bloody Polish uprising of January 1863, when Russian authorities restricted Polish professors from teaching physics and chemistry. Marie Curie wrote that her father was cheated out of what might have been an outstanding career in science by Russian repression. Although Wladyslaw continued to read scientific journals and reports, "my father had no laboratory and could not perform experiments." Certainly, she would not be the first child driven to fulfill a father's thwarted dreams.

Professor Sklodowski led a precarious existence as an underinspector and teacher in a Russian government-run gymnasium for boys in Warsaw. These Russian gymnasia were the only schools authorized to give diplomas. Many of the Polish teachers in these schools were considered by their fellow countrymen to be "polluted" by the Russians. But secretly, Wladyslaw felt that he could keep Polish nationalism and culture alive through his teaching.

Once Poland had been a proud land, but after Napoleon's final defeat at Waterloo in 1815 at the Congress of Vienna, Tzar Alexander II of Russia was named "King of Poland" and this country came under the joint control of Russia, Prussia, and Austria. Even the name *Poland* was expunged from many maps; it was now referred to as "the Vistula," after the river of the same name. The Russians were particularly harsh. The Polish language was forbidden in schools, as was the teaching of Polish history and literature. The official language was Russian, and all street and shop signs were written in Cyrillic.

Two uprisings against the Russian occupation had failed. For the Sklodowski family both hit close to home. In the first, in November of 1830, Wladyslaw's father, Jozef, a respected physics and chemistry professor, fought in the artillery. Captured by the Russians, he was forced to march barefoot one hundred and forty miles to a prison camp. Along the way he lost forty pounds. His feet, bloody and swollen, pained him for the rest of his life. Miraculously, he escaped.

The uprising of January 1863 was an even worse disaster. For a year and a half, Polish fighters—some armed only with spades, clubs, and hoes—met the Tzar's army. By the end, thousands of Polish resisters were dead or exiled to Siberia. One of Manya's uncles was wounded in the fighting. Another uncle spent four years in Siberia. About one hundred thousand Polish resisters took what belongings they could carry and escaped to other countries, mainly France. In August of 1864, the leaders of the insurrection were captured and hanged. Their bodies dangled from the ramparts of the Alexander Citadel a few blocks from the Sklodowski home. The corpses were left all summer to rot in the heat.

Professor Sklodowski fought the battle from within. Like many intellectuals, he realized that open rebellion was fruitless. In 1860, while the insurrection against the Tzar was beginning to ferment, at twenty-eight he married a beautiful, accomplished young woman, Bronislava Boguski. Both were from the lower aristocracy known as the Szlachta. Though this class had managed to keep some trappings of aristocracy, such as royal crests and villages that bore their family names, over the years most of them had lost both their lands and wealth. However, they retained their love of learning, becoming priests, doctors, teachers, musicians. Some 40 percent of the peasant class were richer, but the Szlachta, steeped in the memory of faded glory and intellectual achievement, felt vastly superior to those who measured themselves in worldly goods.

Jozef Sklodowski, Manya's grandfather, had attended Warsaw University but chose to teach in the less repressive provinces. Her father, too, wanted to attend Warsaw University, but it was temporarily closed after the 1830 rebellion. Wladyslaw was forced to take private tutoring in biology and then attended the Science University of St. Petersburg, where he secured a degree in mathematics and physics. He then returned to Warsaw and was hired as an assistant teacher. His salary was so meager that it would not have allowed him to marry. Bronislava Boguski came to the rescue.

The commonly held view at the time that women were neither physically nor mentally equipped to enter the workforce was shattered by a hidden reality: peasant women worked in factories and sweatshops for a fraction of men's wages, and on farms tilled the soil and harvested crops. During the 1863 rebellion, it was women who took over men's jobs with great efficiency. After the insurrection failed, women, often reluctantly, were once again relegated to marriage, childbearing, and household duties. The professions open to them were limited, teaching and nursing being the most common.

Bronislava's parents were not wealthy but had managed to send her to the Freta Street School, the only private school for girls in Warsaw. All such private schools were monitored by Russian officials, but their scrutiny of this girls' school was not as intense as in comparable boys' schools since Russian officials believed that women would never enter public life, politics, or indeed any other influential position in a man's world.

By the time Bronislava married Wladyslaw in 1860, she had, by sheer dint of intelligence and exceptional scholastic ability, worked her way up from being a teacher to headmistress of the Freta Street School. She had a steady income and use of a spacious ground-floor apartment adjacent to a wing of the school. With marriage, Bronislava began to lead the life typical of a woman of her time, but she also assumed the added burden of providing financial support. In the next six years she bore five children: Zofia (nicknamed Zosia) in 1862; Jozef in 1863; Bronislava (Bronya) in 1865; Helena (Hela) in 1866; and on November 7, 1867, the same year that Karl Marx published the first volume of *Das Kapital* and Alfred Nobel patented dynamite, her last child, Marya Salomee (Manya). After all this, Bronislava told a friend, "I must confess that I wouldn't mind being Miss Boguski again, now that I see how difficult a woman's life is."

It was in 1867 that Bronislava's husband was appointed assistant director of a Russian gymnasium on Novolipki Street. His new position came with an apartment. There was no question that Professor Sklodowski's career would take precedence over that of his wife. The family, four daughters and a son, promptly moved from the center of Warsaw to its western outskirts. For a short time Bronislava began the long commute to the Freta Street School, but with her children no longer nearby, coupled with the strain of her other responsibilities, her health began to suffer. She resigned and became a full-time housekeeper, tutoring Zosia and Jozef at home. In order to save a few rubles she also taught herself to be a cobbler and set up a bench to make her children's shoes for the cost of the leather only. The *tap-tap-tap* of her hammer was the accompaniment to their childhood lessons.

In 1871, when Manya was four, her mother began to lose weight. She coughed constantly, a classic sign of tuberculosis. Manya was never to remember a mother's kiss or caress. No doubt this was a precaution her mother took, along with using her own set of dishes and separate eating utensils, but the little girl, yearning for affection, painfully felt the distance. The mores of the time also caused an abyss between parents and children. Her parents had complete authority and were addressed in a formal manner. Manya did as she was told. She could never ask what was wrong with her mother.

On the advice of two doctors, Wladyslaw, though strapped for money, decided to send his wife away on what was to become a series of cures. Bronislava reluctantly obeyed. The theory at the time was that tuberculosis could be cured by an extended stay in a mild climate or in the mountains, by resting, and by imbibing curative waters. It would be nine years before the tubercle bacillus was isolated and people began to realize that it was a contagious disease. Unable to afford a nurse, she took her then ten-year-old daughter Zosia with her. This child pathetically tried to take care of her mother as an adult nurse might.

As the separations grew longer, Bronislava became dispirited. A cure in the Austrian Alps near Innsbruck was followed by one in Nice. Bronislava worried about the money her treatment was costing. As the time stretched into a second year, she and Zosia became more and more homesick. In Nice on Christmas Eve, Zosia set the table as she had at home, and these two with tears in their eyes broke the sacred wafer that had been sent from Warsaw. "May God make this the . . . last Christmas away from my family," Bronislava prayed.

In the absence of his wife, Professor Sklodowski took over

the care of his other children, and circumstances were to dictate that this would remain so for the rest of their upbringing. This man in his shabby black coat became the supreme commander of his small troop. The children's days and evenings were carefully parceled out into periods of study and exercise. Manya recalled that even the most casual conversation contained a moral or academic lesson, that a walk in the countryside served as a means to explain a scientific phenomenon or the mysteries of nature and a sunset provided a speech on astrological movements. Since their mother was an ardent Catholic, the children learned their catechism and an aunt took them to church every Sunday, where they prayed for their mother's return. At home they were instructed to add to their evening prayer, "restore our mother's health."

Wladyslaw instilled in his children the hope of Polish nationalism and a deep hatred of Tzarist Russia. On her way to school Manya and a friend would stop in front of an obelisk, erected by Tzar Alexander II near Saxony Square, which bore the inscription TO THE POLES FAITHFUL TO THEIR SOVEREIGN. Taking aim, she would spit on those hateful words. When the Tzar was assassinated by a terrorist bomb in St. Petersburg, Manya and her Polish classmates were found dancing for joy in the schoolroom.

In the regimented lives of the children, Saturday night provided a pleasant interlude. From seven to nine, their father— who was fluent not only in his native language but also in Russian, French, German, and English—would read aloud such books as *David Copperfield,* translating the text into Polish as he went along. Manya was particularly touched by *A Tale of Two Cities*, in which a patriot had been reduced to making shoes, just like her mother.

All the Sklodowski children were bright and excelled in school, but Manya was the most brilliant. At four, observing her elder sister Bronya's struggle to read, she picked up Bronya's book and read aloud the first sentence flawlessly. Then, noticing the astonished faces around her, she began to cry, thinking she had committed an unpardonable sin. "I didn't mean to do it," she whimpered plaintively, "but it was so easy." Several years later an acquaintance read her a poem and she asked for a copy of it. Teasing her, he said he would read it once again and since her memory was supposedly so good she could then, no doubt, recite it by heart. He read the poem. Manya went into another room and emerged half an hour later with a perfect written transcription.

Manya and her sisters had originally attended the Freta Street School, but when she was six and a half she and Helena were transferred to a school closer to home. Manya was placed in third grade although many of her classmates were a year to two older. This school was more closely monitored by Russian officials than her former school, but it was run by a Polish patriot, Madame Jadwiga Sikorska, who in order to deceive the officials secretly wrote a double schedule. Her students understood that Polish history was on the schedule as "Botany"and Polish literature as "German Studies." A skillful system was devised whereby if a Russian official approached, a bell would ring and the Polish books would vanish from the classroom while Russian ones took their place. Once, when an inspector arrived, Manya, the star pupil, was selected to answer his questions in her perfect Russian. The last question he asked was, "And who is our beloved Tzar?" Manya paused, then answered in a choked voice, "Tzar Alexander II." As soon as the door closed she burst into tears at her own perfidy. But she had begun to learn that showing her real feelings could produce disastrous results.

Her father, whom she worshipped, also led a double life, secretly lecturing on Polish scientists to instill pride in his students' heritage. At the Novolipki Street School the head inspector, who was Russian, discovered Professor Sklodowski's subversive activity. He was abruptly fired, losing both his apartment and salary just as Bronislava determined that no matter how sick she was she would return home. When Manya saw her mother and eldest sister once again, she ran into Zosia's arms, but her mother held up her hand, palm extended, so that Manya would not come near. The six-year-old child stopped short, hardly recognizing this wraith of a woman with a hacking cough. That Sunday in church Manya knelt and prayed. She told God she would give up her own life if her mother were cured.

The family now rented a house, and to pay the upkeep Professor Sklodowski opened a boys' boarding school, mostly for students from the provinces. At first there were five, then ten, then twenty. In these quarters there was little privacy. Manya slept on a couch in the dining room. Every morning she rose at six to set the table for breakfast.

In January of 1874, one of the many boarders infected both Bronya and Zosia with typhus. Transmitted by lice and rat fleas lodged in dirty clothes, bedding, or fur, typhus flourished in overcrowded conditions. Two previous epidemics in Warsaw had killed thousands of people. Manya's sisters lay shaking with fever while in the next room her mother's cough could be heard day and night. After twelve days, Bronya recovered. Two weeks later, twelve-year-old Zosia, her mother's dear companion, died. Bronislava, too ill to go to the cemetery, stood at the window as the cortège moved by. Manya, dressed in her dead sister's long black coat, walked behind the coffin as if in a trance. In May of 1878, Bronislava finally succumbed to tuberculosis. Manya was to write that her mother at forty-two had been "cruelly struck by the loss of her daughter and worn away by grave illness." The following Sunday Manya went to church as usual, but as she knelt she reflected that never again would she believe in the benevolence of God.

The pain of these two deaths expressed itself in what she called "a profound depression," the beginning of a pattern that would remain with her all her life. Later, when she became Madame Curie and world attention was focused on her, she became less frank, calling these episodes "fatigue" or "exhaustion" or "my nervous troubles." Today experts would diagnose her condition as a recurring major depressive disorder which is often triggered by grief or loss. It was months before she stopped creeping into deserted spaces and crying, but she hid this from her family and schoolmates. She carried on with her schoolwork with no sign of grief and remained at the top of her class. Soon after her mother's death, Manya seemed to lose herself in books for hours, sometimes days, at a time. She spoke little. The only way she was able to cope was by screening out the world and focusing obsessively on a subject, thus holding at bay her feeling of desolation. Years later, Eve remembered coming home at 3 A.M. and seeing the light on in her mother's room. Upon entering she found her mother poring over scientific papers unaware of her daughter's presence. From childhood, depression and withdrawal marked Manya and the adult Madame Curie she was to become.

At the end of the 1879 school year Madame Sikorska, the headmistress of Manya's school, visited Professor Sklodowski and informed him that, although Manya was ahead of herself in school, she was unusually sensitive and mentally fragile. She suggested that he wait a year before letting her begin the next grade. Her father did just the opposite. Since Russian-run schools were the only ones that led to higher education, he removed his daughter from Sikorska's nurturing environment and enrolled her in Russian Gymnasium Number Three. The level of education at the gymnasium, which once had been German, was excellent, but the Russian effort to expunge Polish culture was emotionally disturbing. In her years there Manya keenly felt that the teachers treated the Polish students as enemies. As a young child she had written that sometimes, when she felt angry or deserted or forced to lie, she wanted "to scratch like a cat," but now she rebelled in more subtle ways. When one of her teachers reprimanded her for her superior attitude saying, "I feel you look down on me," Manya, who was taller than her teacher, answered, with anger disguised as humor, "The fact is that I can't do anything else."

One by one the Sklodowski children fulfilled their father's expectations by graduating first in their class with all the attendant honors, with the exception of Helena, who received a second place. She felt despondent that she had failed her father. Manya Salomee Sklodowska graduated the government gymnasium with a first in her class and a gold medal for the best pupil of 1883. She was fifteen.

After years of pressure, of performing perfectly, of deception, of suppressing her passionate feelings, and of despair, she had a total nervous collapse. She took to her bed in a darkened room, would not speak, and ate little. Her father, finally alarmed, decided to send her to relatives in the country to regain her health and equilibrium. And thus began what was to become the happiest, most perfect year of her life.

The Boguskis and the Sklodowskis were part of an extended family, and a few of their relatives had managed to retain their manor houses and some of their wealth. Manya spent the first part of the summer in the south at the home of a Boguski uncle. At first she was so debilitated and depressed that she could only rest, but then she began to recover her good spirits. Manya abandoned her science books and read novels, fished and gathered wild strawberries with her cousins, took long walks, rolled hoops, played shuttlecock and tag, and enjoyed "many equally childish things." She was given a sketchbook, in which she displayed both talent and humor. One sketch was of the family dog eating from her dinner plate. Manya wrote, "Sometimes I laugh all by myself and I contemplate my state of total stupidity with genuine satisfaction." She was having the childhood she had never experienced.

In November, she visited another uncle who lived further south in the foothills of the Carpathians. Her uncle and a cousin both were talented violinists. This too was a happy household filled with music, books, and art. Then, just when it seemed the fun would end, one of her late mother's former pupils who had married well invited Manya and Helena to her country estate northeast of Warsaw. The parties there were even more elaborate than those of Manya's uncles, and Helena recalled that the time "passed as quickly as a dream but the memory of it has been lasting."

Years later Marie told her daughter Eve of that magical year, where her uncles and aunts had lavished gifts on her, where sleighs filled with laughing young people set out at night going from one manor house to another, feasting, flirting, dancing the last mazurka at dawn. She told Eve how one night she had danced so long that she had tossed her slippers in the trash, for there was nothing left of them. Eve, who was born when her mother was thirty-seven and was only fourteen months old when her father died, could hardly imagine—in the dour, silent, isolated, seemingly impassive mother that Madame Curie had become—the happy, laughing, open young girl who had danced all night.

CHAPTER 2

"I Came Through It All Honestly"

At sixteen Manya returned to Warsaw. Her perfect posture and porcelain skin set off by intense grey eyes revealed the beauty she was to become. Professor Sklodowski had given up boarders, taken a lesser job, and moved to a smaller apartment. Though drab and uncomfortable this offered privacy. Although Manya's father professed to regard education as genderless, nevertheless what little money he had went to paying for Jozef's medical education. In any case, Warsaw University banned women. Still Manya and her older sister Bronya had dreams. Big dreams. At eighteen Bronya had assumed the role of her mother but longed to be a physician like her brother. Manya wanted to be a scientist or at least to become "something," by which she meant a person of importance to the world. She continued to educate herself and read science, politics, and literature. With her mother's death, she had abandoned much of her belief in religion and copied into a schoolgirl notebook a passage from

Max Nordau's attack on the institutional deceit within the church. When a cousin's child was stillborn, she wrote,

> If one could only say, with Christian resignation, "God willed it and his will be done!" half of the terrible bitterness would be gone. . . . I see that happy people are those who believe such explanations. But, strangely enough, the more I recognize how lucky they are, the less I can understand their faith, and the less I feel capable of sharing their happiness. So far as I am concerned, I should never voluntarily contribute toward anybody's loss of faith. Let everybody keep his own faith so long as it is sincere. Only hypocrisy irritates me—and it is as widespread as true faith is rare. . . . I hate hypocrisy.

Manya had loved Poland with the fire of a child, but now her views had cooled to a more intellectual stance. Auguste Comte, a French philosopher who had lived through the chaos that followed the French Revolution and Napoleon's reign, introduced the term *positivism* in response to the then popular abstract study of classical philosophy. Science and technology were beginning to transform society, and Comte rejected the theoretical in favor of a new positive philosophy that proposed improving society by imposing methods that could be verified by empirical observation. His philosophy called for governing groups to guide people to a better future. He believed that bettering the education and moral consciousness of a person would better society itself.

After Comte died of cancer in 1857, other philosophers bent his strict views to suit their own needs. In Poland, positivism took the form of opposition to church strictures.

Though Comte had not supported women's rights, sexual equality, and the emancipation of women, Polish positivists espoused these causes and found in Comte's philosophy a way to assert nationalism without unnecessary bloodshed. They urged the education of workers and peasants in Polish traditions, language, and history, thereby keeping the flame of nationalism alive until the Russian oppressors could be expelled. Polish positivists recognized that this nonviolent approach might take years and advocated patience and dedication to reach their goal. All this appealed to Manya, who later wrote, "I still believe that the [positivist] ideas which inspired us then are the only way to real social progress. You cannot hope to build a better world without improving the individuals." Also, in the scientific career that was to follow, she believed with the positivists that all statements and conclusions should be "supported by evidence which can be checked." This belief replaced religion in her life and became one of the keys to her success.

The year Manya graduated from the gymnasium, a Polish positivist started a clandestine academy for the higher education of women and within a year had enrolled over two hundred women who met in secret. After a few months, they were discovered by the Russians and most of the teachers were exiled. This served as a challenge. In the next three years the academy became known as "the Flying University," with over a thousand women enrolled, Manya and Bronya among them. In their classes they met other like-minded women, including several who had been widowed in the 1863 uprising and now responsibly ran family estates and businesses, as well as women who valued higher education and envisioned going abroad to universities that accepted women. One of the courses was taught in the home of Manya's sympathetic former headmistress, Jadwiga Sikorska. Others were in well-known institutions around Warsaw. The Russian authorities must have been aware of this, but the academy was now too large to forcibly eliminate without embarrassment. And what could women do anyway?

Bronya and Manya knew that they would have to be self-supporting. They became tutors while they continued to study at the Flying University. They began by tutoring at home or walking long distances across Warsaw to teach reluctant or lazy students at the rate of half a ruble an hour. In the first year they earned so little that Manya sought a position as a governess in Warsaw. She was engaged by a newly rich family who spent freely on displays of wealth but were abusive and miserly with their servants. As a child of the Szlachta class, Manya retained her sense of self-worth and intellectual superiority. Her employers found her arrogant. With ironic humor she wrote that the mistress of the household "was exactly as enthusiastic about me as I was about her. We understood each other marvelously well." She also noted, "One must not enter into contact with people who have been demoralized by wealth." Three months later she resigned. "I could not endure it any longer."

At this point the sisters' plans of one day becoming a doctor and a scientist could have faded away had it not been for Manya's inventiveness, generosity, and determination. Bronya had scarcely saved enough to pay for even one year of the required five to obtain her medical degree at the Sorbonne in Paris. Manya told her sister matter-of-factly that she intended to find a job as a governess in the provinces where room and board were free and then send half her salary to Bronya. Their father would contribute what little he could. Then, when Bronya became a doctor, she could help bring Manya to Paris to attend the Sorbonne. Bronya burst into tears, "Why should I go first?" she asked. Manya replied, "Because I am seventeen [almost eighteen] and you are going to be twenty." The following week, wasting no time, Manya went to work for the Zorawski family for 500 rubles a year. They lived some fifty miles north of Warsaw in Szczuki. Expecting a country manor house, Manya found instead that the Szczuki home was situated next to a sugar-beet factory with a tall chimney that spouted black smoke. The Zorawskis administrated the estate of a wealthy family and supervised the peasants who grew and processed the sugar beets. Unlike her first job, this one began well. The eldest daughter, Bronka (another variation on her late mother's name, Bronislava) was almost nineteen, only a year older than she was. There was a child of ten, Andzia; a six year old, Maryshna; and Stas, a boy of three. The eldest son, Casimir, was studying mathematics at Warsaw University.

At first Manya was welcomed into the family and treated as a surrogate daughter. She was invited to many of the same social events attended by Bronka, but Manya felt like a poor outsider. She could not afford the pretty clothes that Bronka wore, and her work kept her busy from dawn to dusk. She was shy and ill at ease with new people. She had no taste for small talk and masked her insecurity with an air of intellectual superiority. This led to trouble. Soon after her arrival she wrote, "They were already speaking of me unfavorably because, as I didn't know anybody, I refused to go to the ball." The previous year with her relatives she had relished dancing, parties, and "childish things," but now she wrote home that similar events held no interest for her. She noted that the local girls danced perfectly (an attribute she had shared the previous year) but that their "stupid incessant parties here have ended by frittering their wits away." Her change in attitude most certainly was based on her change in social status. A prominent author of the day wrote that a governess was certainly a lady, but a needy one: "There is no other class which so cruelly requires its members to be, in birth, mind, and manners, above their station in order to fit them for their station."

Manya wrote, "In a general way I observe, in my talk, the decorum suitable to my position." But she imagined a better future. At night she studied and rose before six to continue her self-education. She wrote that she was reading "Daniel's *Physics*, of which I have finished the first volume; Herbert Spencer's *Sociology* in French; Paul Bers's *Lessons on Anatomy and Physiology*, in Russian." She gravitated toward mathematics and physics and struggled alone to learn the material she felt she needed to know for the day when Bronya would send for her. Her father consistently sent her math problems to solve and cautioned her to keep studying so she would not fall behind in life. At eighteen, she had already "acquired the habit of independent work": to draw her own conclusions, without the restraints of accepted perceptions, later would help her in her amazing discoveries.

Manya's one real friend in the Zorawski household was Bronka, who, inspired by Manya's desire to serve others, agreed to help her start a forbidden project: teaching the illiterate peasant children on the estate how to read and write in Polish. This too was considered a crime by the Russian government and meant exile to Siberia, but these two young women were unafraid. Eventually, twenty shy, dirty, but eager children, most accompanied by apprehensive parents, attended night classes in the kitchen of the Zorawski house.

In the spring Casimir returned from Warsaw University for vacation to find a beguiling governess in residence. She understood mathematics and didn't gossip, she spoke and wrote three languages fluently, and like Casimir she loved nature. Soon it was clear that at eighteen Manya was in love. The first clue came in the form of a protestation. She wrote a friend, "Some people pretend that in spite of everything I am obliged to pass through the kind of fever called love. This absolutely does not enter into my plans." But in love she was, and by late summer the young couple told the Zorawskis of their plan to marry. Abruptly, any pretense that Manya was an equal was gone. Casimir's father flew into a rage, insisting that his son would never be permitted to marry a penniless governess who was obliged to work in "other people's houses." His mother was equally appalled and pointed out that if he pursued this reckless course he would be disinherited. Casimir begged Manya for patience; he would work it out. He would defy his parents but retreated when he realized that he would be unable to pursue his education without their help and would lose all social status in the community if he married "below his station." For her part, Manya, though humiliated, rationalized that she could not leave because she needed to send money to Bronya until her sister completed medical school.

Her letters at this time reflect her ever-darkening mood and the recurring depression that plagued and isolated her. She flagellated herself with a lack of self-worth. She wrote of feeling "stupid" and as poor as the Zorawskis said she was. "I have literally not a ruble—not one," she wrote her brother, Jozef. Of her studies she complained, "What can I do as I have no place to make experiments or to do practical work?" Her complaints echoed those of her father so many years before. With each letter she seemed more despondent. "If you only knew how I sigh and long to go to Warsaw for only a few days. To say nothing of my clothes which are worn out and need care—but my soul too is worn out. Ah, if I could extract myself for a few days from this icy atmosphere of criticism." At Christmas Casimir returned from the university. Once more he braved his parents to no avail. As Edward Gibbon wrote, when told by his father that he could not marry a Catholic, "I sighed as a lover. I obeyed as a son."

Manya said nothing, but nonetheless revealed her anger when her beautiful sister Helena received similar treatment from a rich suitor.

> I can imagine how Hela's self-respect must have suffered. Truly it gives one a good opinion of men! If they don't want to marry poor young girls, let them go to the devil! Nobody is asking them for anything. But why do they offend by troubling the peace of an innocent creature? . . . Even I keep a sort of hope that I shall not disappear completely into nothingness.

But "nothingness" was descending and she wrote, "I have fallen into black melancholy."

Unexpectedly, a reprieve from this untenable situation came from her father, who wrote that he had accepted a position as director of a reform school outside Warsaw. The job was repellent but the salary was excellent, and after two years he would receive a pension that would let him live in modest comfort for the rest of his life. He told his daughter to send no more money to Bronya, as he would assume that obligation. Manya gave notice to the Zorawski family. She left with a smile on her face. She had learned well the lesson of dissimulation:

> For me I am very gay—for often I hide my deep lack of gaiety under laughter. This is something that I learned to do when I found out that creatures who feel as keenly as I do and are unable to change this characteristic of their nature have to dissimulate it at least as much as possible.... There were some very hard days and the only thing that softens the memory of them is that in spite of everything I came through it all honestly with my head high.

Manya returned home. Bronya had graduated from the Faculty of Medicine, one of only three women in a student body of a thousand. She had met a Polish émigré doctor, Casimir Dluski, who had fled Poland because of his socialist views. He disagreed with the positivists and in his earlier days had written a diatribe against them, saying that socialism was a greater form of patriotism and that only this philosophy would free Poland. In spite of their political differences, they were very much in love. Casimir could not be married in Warsaw, so both families assembled in the Austrian-ruled Krakow, where the regulations were so lenient that no one was penalized for their political views and one could sing the Polish anthem with impunity. Shortly after the Dluskis returned to Paris, Bronya wrote that if her sister could secure a few hundred rubles she could live with them and in two years acquire the degree at the Sorbonne that she so coveted. But after so long a time Manya's dream had all but vanished. It seemed

Wladyslaw Sklodowski with his daughters Manya, Hela, and Bronya in 1890.

that she had lost both her focus and her courage. She answered:

> Dear Bronya, I have been stupid, I am stupid and I shall remain stupid all the days of life. . . . I have never been, am not now, and shall never be lucky. I dreamed of Paris as a redemption but the hope of going there left me a long time

ago and now that the possibility is offered me I don't know what to do. . . . I am exceptionally unhappy in this world.

By now Wladyslaw was well aware of his daughter's periods of severe depression and deduced that she was still hanging on to the idea that Casimir Zorawski would marry her. He wrote Bronya that "the strain of this has added to her disorder." The following summer Manya announced that she was meeting Casimir at a resort in the Tatra Mountains and confided to her father that she had "a secret about [my] future." Wladyslaw then wrote Bronya, "To tell the truth I can well imagine what it has to do with and I don't myself know whether I should be glad or sorry. If my foresight is accurate the same disappointments, coming from the same persons who have already caused them to her, are awaiting Manya." But, with hope as unrealistic as that of his daughter he added, "How funny it would be if each of you had a Casimir." What was left unsaid and would have been unthinkable to Professor Sklodowski was that a wealthy man who desired a woman below his station might persuade her to become his mistress. What happened between Manya and Casimir is unknowable, but had the question come up one can imagine her answer. On this trip, Manya broke with Casimir telling him, "If you can't see a way to clear up our situation it is not for me to teach it to you." After this, Manya began to recover from her bout of depression. She confided to a friend,

Everybody says I have changed a great deal physically and spiritually during my stay at Szczuki. This is not surprising. I was barely eighteen when I came there and what I have not been through! There have been moments which I shall certainly count as the most cruel of my life. . . . I feel everything very violently . . . with a physical violence and then I give myself a shaking, the vigor of my nature conquers and it seems to me that I am coming out of a nightmare. . . . First principle: never to let oneself be beaten down by persons or by events.

In September of 1891, she wrote to her sister that if Bronya could give her room and board without deprivation to herself, she would register at the Sorbonne. After almost eight years of work, Paris and a new life were about to begin.

CHAPTER 3

Paris

After one last tearful caress Manya assured her father that when she received her degree in science she would return to Warsaw to live with him and teach the next generation of Polish patriots. At the end of November 1891, she bundled up her clothes, feather mattress (who knew how much they would cost in Paris?), food, water, and a stool. She bought the cheapest train ticket to Paris and began a thousand-mile journey into the unknown.

At twenty-three Manya's character had been formed. She had learned that if she had enough patience and tenacity the seemingly impossible could be accomplished. She masked her feelings with a cool intellect. "We should be interested in things not persons," she later wrote, indicating how she had coped with emotional deprivation. All this enabled her to ignore such obstacles as sexual discrimination, lack of money, and her inadequate preparation in chemistry and physics. For this intellectually ravenous young woman, the Sorbonne promised a feast. Casimir Zorawski had cost her four heartbreaking years and solidified her tendency to beware of personal relationships. Now she would devote herself to science.

After almost four days of sitting on her stool day and night and rationing her food, Manya descended into the Gare du Nord station. Bronya's husband, Casimir Dluski, was awaiting her and took her to their apartment on the rue d'Allemagne in a working-class neighborhood, a short distance from the station. During the day the apartment was used as medical offices by Bronya and Casimir. Casimir saw both male and female patients, but in keeping with the practices of the time women kept their clothes on, making diagnosis difficult. Bronya saw only women patients and was allowed to examine them more freely. Also, in accordance with Casimir's socialist views, two evenings a week, patients were allowed to come free of charge.

Within the week Bronya took her sister to register for classes at the Sorbonne. With a newly found "very precious sense of liberty and independence" she signed the registration forms not with her Polish name, Manya, but with the French equivalent, Marie. For someone accustomed to the regimented structure of the Russian curriculum, freedom bewildered her. At the Sorbonne, students could attend whatever classes they wanted, whenever they wanted. Exams were voluntary and could be taken at any time. And all this, with some of the best professors in the world, was free. She was determined to use the weeks before school began to study math and perfect her French. (Her Polish accent drew smiles or glances of derision from the Parisians.) The atmosphere at the Dluskis', however, was lively and distracted her from her studies. She resented it. In Marie's view, in addition to the noisy office hours, Casimir was constantly interrupting her with what she termed his "idle chatter." In the evenings the expansive Dluskis liked to entertain Polish émigré artists, musicians, scientists, socialists, and physicians who talked about home in Polish, which Marie felt interfered with her French.

Once classes started it was even more difficult to study. Her sister's efforts to get her to eat regular meals irritated Marie, and the hour-long omnibus ride in each direction was another obstacle. She found she stayed late at the Sorbonne to avoid the distractions at the Dluskis' apartment. After a few months, Marie took what little money she had and, with a contribution from Bronya, for only twenty-five francs a month rented a sixth-story, unheated garret room at 3 rue Flatters in the Latin Quarter. It was the first of four such rooms that she would rent during the next two and half years. As soon as each semester was over, she would move out to save money. These rooms, former servants' quarters, were occupied mainly by impoverished artists, prostitutes, factory workers, and students who had the advantage of living only a short distance from the Sorbonne. Marie's garret rooms were to become part of her legend.

At night the Latin Quarter seethed with life and frenetic gaiety that often turned violent. Once, while Marie lived there, some thirty thousand troops were called in to quell demonstrators who rioted after the police banned nudity at an artists' ball in Montmartre. Marie, with her high forehead and cheekbones, and unruly blond hair, must have been attractive, but her solemn abstracted innocence provided its own protection as she walked home, invisible in her black wool coat. Late at night she moved past the rowdy students and prostitutes who filled the cafés, past the brightly lit windows where half-naked whores displayed themselves, down the dark streets and up the six flights of rickety stairs to a bare room with its small stove. In her short autobiography written in 1923, in which she helped bolster her own myth in order to finance her scientific research, she wrote that the room was so cold in winter that the water froze in the basin. She would sleep beneath all her clothes piled on the bed. Sometimes she bought a scuttle of coal and hauled it up the stairs, but not often. Sometimes she varied her diet of tea, chocolate, bread, and fruit with an occasional egg or meat, but not often. However, each time she recalled these two and a half years of "deprivation," she pronounced them "one of the best memories of my life." Marie had achieved her dream of study, of liberty, of independence. Here in these "solitary years" she studied science and spoke aloud to the cracked walls in French, still trying for a perfect accent. Marie wrote home that she was getting a thousand times more work done than she had at the Dluskis'.

One of only twenty-three women among almost two thousand students enrolled in the School of Sciences, she made no comment on this disparity but noted only that she was being taught by such illustrious professors as Paul Appell, who was to become dean of that institution, and Gabriel Lippmann, who in 1908 would win a Nobel Prize for developing color photography. Lippmann along with the two Curie brothers, Pierre and Jacques, had designed several measuring devices that were being used by other physicists. The most famous of Marie's professors was Henri Poincaré, the great mathematician, physicist, and philosopher whose research into celestial mechanics was of fundamental importance in understanding our solar system.

In the tradition of French gentility, in class Marie was treated with respect if not given the same attention as the male students. When she stepped outside the doors of the Sorbonne, she entered a different world. The condition of women in Poland and France was similar in certain ways: Women were taxed if they owned property but they had no voice in politics. Divorce, then rare, meant giving up all rights to property, income, and child custody. Neither sanctioned by church or state, it meant poverty for most women. Men had no obligations in this area. If a woman ran away, a husband could legally track her down and return her as if she were stolen goods. There were no laws to prevent the abuse of wives or children. In France the condition of women was influenced a great deal by Gallic sexual attitudes. The word for a female student, *étudiante*, was also used for the mistress of a male student at the Sorbonne. In a country where well-bred women never ventured out unchaperoned, never went to a restaurant alone, never received a gentleman alone in their rooms, Marie seemed oblivious to these restrictions:

> All my mind was centered on my studies. I divided my time between courses, experimental work, and study in the library. In the evening I worked in my room, sometimes very late into the night. All that I saw and learned was a new delight to me. It was like a new world open to me, the world of science which I was at last permitted to know in all liberty.

Marie was unconcerned or unaware that she belonged to the "weaker sex"; a popular book of the day was titled *The Physiological Feeble-Mindedness of Woman*. Among the upper classes, a good way to rid oneself of an independent-minded wife was to confine her to an insane asylum. A French critic wrote that "women's place is simply for sex and to reproduce." The rare female scientist was depicted as masculine, coarse, ugly, careworn, and industrious but making no significant contribution. The brightest of these women served as invisible assistants to their presumably superior male counterparts.

A double standard was firmly in place: A man's prestige was enhanced by his bejeweled mistresses. The literature of the time celebrated male conquests. Conversely "loose women" were stigmatized. Upper-class women were allowed discreet affairs but to reveal them meant ostracism. Flaubert's *Madame Bovary,* though written in 1856, still provided the prototype of a woman foolish enough to fall in love with her lover. Abandoned by her first lover, she squanders her money on a passionate second affair. Finally, realizing that her husband will discover her actions, she commits suicide by drinking arsenic and dies a miserable death. Implicitly, these are the wages of violating the accepted code.

In *Anna Karenina* the message is the same. Anna, a beautiful young woman married to a high government minister, has an affair and leaves her husband and son to live with her lover. Her lover, Count Vronsky, a dashing young military officer, enjoys social freedom, but Anna is shunned by a society that once glorified her. Her increasing suspicion sours Vronsky's love for her. Tortured by her own jealousy and unable to bear the loss of her son, Anna throws herself before an oncoming train. The story of Anna and Vronsky is contrasted against that of a happy young couple who marry, move to the country, have a son, and enjoy family life. This, Tolstoy declares, gives meaning to life and advances the will of God.

But none of this affected the obsessed student. Like the parched earth, she soaked up education as a life-giving force. In what was to become an enduring pattern, when she studied, the world around her vanished. She seemed to live on air. A fellow student advised her to make soup to keep up her strength, but she had no idea how to do it and did not want to squander her time learning. She did not know where to shop and, as if there were no tea or an iron in all of France, she wrote her father asking him to send these items from Warsaw. One day Marie fainted in the library and a Polish student told Bronya of this. Casimir took the protesting Marie to the Dluskis' apartment. Bronya put her to bed and fed her steak and potatoes, recognizing that her disease was simply fatigue and malnutrition. As soon as she regained her health, she rushed back to her garret and her studies.

As summer drew near, Marie found herself one of only two women pursuing a degree in science. She seemed unaware of her own academic excellence, writing, "The nearer the examinations come the more I am afraid of not being ready." In July, in the examination hall, she was so nervous when the examination was handed to her that she could not read it for several minutes. Then she forced herself to be calm and took the exam. The results were not published for several days. When they were announced, Marie slipped in among the crowd in the Sorbonne's great amphitheater to hear the names of those who had passed read in order of merit. The examiner entered the room and in an atmosphere of apprehensive silence began to read the list: The first name was Marie Sklodowska.

Money was running out. With twenty-five rubles for rent, she had fifteen rubles a month for everything else. But even then, her efforts were already becoming an inspiration to women in Poland, and one used her influence to see that Marie was awarded the Alexandrovitch Scholarship of 600 rubles presented to talented students who wished to study abroad. Marie now had enough money for another fifteen months in Paris. In July 1894, she took the exam for her degree in mathematics and ranked second in the entire class. She chastised herself for not ranking first.

Soon after, Marie had unexpected good luck: Her professor, Gabriel Lippmann, with paternal largesse arranged for the Society of the Encouragement of National Industry to pay his impoverished young student 600 francs to study and chart the magnetic properties of various steels. Marie promptly set up her equipment in Professor Lippmann's laboratory at the Sorbonne, but the equipment was so cumbersome and her space so limited that she made little progress. A friend of Bronya's, hearing of her plight, recommended that she meet a little-known physicist who was nevertheless France's most eminent expert on the laws of magnetism with which she was working. More importantly, Marie knew when she heard the name Pierre Curie that he had invented a number of delicate instruments that might help with her work.

CHAPTER 4

Pierre

"Upon entering the room I perceived, standing framed by the French window opening on the balcony, a tall young man with auburn hair and large limpid eyes. I noticed the grave and gentle expression of his face, as well as a certain detachment in his attitude, suggesting the dreamer absorbed in his reflections. He showed me a simple cordiality and seemed to me very sympathetic. After that first meeting, he expressed the desire to see me again and to continue our conversation of that evening on scientific and social subjects, in which he and I were both interested, on which we seemed to have similar opinions."

Written thirty years after her first encounter with Pierre Curie, this description reflects in retrospect the man Marie was to come to love. In fact, her first consideration at this meeting was a practical one, to find a laboratory space and the expertise that Pierre Curie might supply. Marie, scalded by her affair with Casimir Zorawski, had vowed never again to let passion triumph over her mission. For his part, Pierre at thirty-four was still living at home. He was uneasy in the company of women and shared the view that they distracted from important work, and by exuding a *vicieuse* sexuality, led men away from high ideals. Pierre Curie needed a calm atmosphere and could only concentrate on one project at a time. At twenty-two he had written, "Women much more than men, love life for life's sake . . . and draw us away from dedication. . . . It is with women we have to struggle, and the struggle is nearly always an unequal one. . . . Women of genius are rare."

In any case, Pierre Curie could not offer Marie any space for, unbeknownst to Bronya's friend, he had no laboratory himself. However, he could offer her expert advice on the use of a state-of-the-art quadrant electrometer that he had perfected with his brother, Jacques. The quadrant electrometer owed its name to its horizontal plates split into four sections (or quadrants). An industrial firm, the Central Society for Chemical Products, sold the Curies' instruments. After using one, William Thomson, Lord Kelvin, who had constructed an earlier electrometer, journeyed to Paris to meet the man who had made such a superior device, only to find Pierre Curie working in "a cubbyhole between the hallway and a student laboratory" at the School of Industrial Physics and Chemistry of the City of Paris (EPCI).

Pierre Curie had already made several other outstanding contributions to science. Shortly before meeting Marie he had formulated a general principle of symmetry known as Curie's Law. ("The coefficient of magnetization of a body feebly magnetized varies in inverse ratio to the absolute temperature.") This law, which is still in use today, had its inception in Pierre's youth, when he became interested in crystals and their electrical properties. Ancient Greeks noticed that amber became electrically charged when vigorous rubbing was applied. (Both Pliny the Elder and Plutarch used the word *elektron,* which meant "amber," to describe this charge.) Charles Friedel, Jacques Curie's professor, observed that asymmetric crystals (their extremities differ) acquire a polar electricity, known as pyroelectricity, when exposed to a difference in temperature on the opposing ends.

In 1880, when Pierre was twenty-one, he and his brother, Jacques, discovered that, as Pierre wrote, "While stretching [similar crystals] we observe the same phenomenon as pyroelectricity but the charges are reversed. The amount of free electricity is proportional to the change in pressure. We have decided to call this phenomenon piezoelectricity." Then the Curie brothers skillfully cut the edges of a quartz crystal into two parallel sides and covered the quartz with two sheets of tin linked to their electrometer. They applied a charge that deformed the crystal. Using this method, Pierre and Jacques discovered that the amount of electrical charge generated by a piezoelectric quartz provided a precise way to balance the very weak currents emitted by the electrometer that otherwise could not be discerned or calibrated.

At this time, the importance of their discovery of piezoelectricity was unknown. The application of the Curies' discovery, however, has led to substantial scientific advances that today we take for granted—sonar, ultrasound, mobile telephones, television tubes, and electrical appliances. The list goes on and on. On our wrists are quartz watches, where a continuous electric voltage applied to the crystal by a battery creates a piezoelectric pressure, which causes the quartz to vibrate and supplies the resonance that makes the watch keep time.

For all his scientific talent, Pierre Curie was self-effacing. The EPCI, where he worked, was not an exalted institution like Paris's École Polytechnique or the École Normale Supérieure (whose graduates were referred to as either Polytechniciens or Normaliens), but an industrial school that trained engineers and chemists. He had been there for over a decade. At twenty-three, when he became the sole instructor in physics, he had been barely older than his students.

The night Pierre met Marie he was unaware of how much they had in common. Pierre had met a "woman of genius" and one who understood "his nature and his soul." Soon, recalling his youthful disdain of women, he wrote, "I am far away these days from the principles I lived by ten years ago." Neither Marie or Pierre were tied to the established conventions of the day: She thought nothing of giving him her address and meeting him alone in her unfurnished room. She pulled out a trunk for him to sit on while they discussed scientific matters and drank tea. Though they were from different countries, their backgrounds were similar: Both came from families of high intelligence and modest means. Pierre's father, Eugène Curie, was a physician, as was his grandfather. Both Marie's and Pierre's fathers had pointed out how scientific observations could be made from what one saw in the natural world, imbuing their children with a sense of the invisible universe that was all around them. Both fathers were patriots, dedicated to the improvement of the common man and to a love of freedom. Both had suffered in failed insurrections. Pierre had been twelve in 1871, when the forces of the Paris Commune and those of the entrenched government clashed. He and his brother, Jacques, had carried the wounded from the barricades back to their father for treatment.

When finally Pierre took Marie to visit his father and mother, she was struck by the similarity to the atmosphere in her own home, based on parental love and family closeness. Marie had received a repressive, clandestine, autodidactic education, but Pierre's background too was unusual. He had attended no lower school. At an early age he was unable to read or write but had an ability to visualize mathematical concepts far beyond his years. His father, unusually enlightened for his time, had realized that his son's spirit would be broken in a regular school. He had decided on home-schooling Pierre, aided by his wife and Jacques. Today, one would diagnose Pierre Curie as dyslexic. His handwriting remained that of a child and his spelling was abominable. He could not concentrate when there was any noise or distraction, but once he embarked on a project he became totally absorbed in it. He was not unlike his contemporary, Albert Einstein, also a poor student, who could not work with interruptions but could visualize spatial concepts.

At fourteen, Pierre developed an attachment to an excellent tutor who taught him mathematics and Latin. By the age of sixteen he had received his science baccalaureate and, unsure whether he wanted a career in chemistry or physics, attended two schools simultaneously, taking a degree in physics at the Sorbonne and enrolling at the School of Pharmacy in Paris, where Jacques had become an assistant chemistry professor. Then his interest shifted once more, and he became an assistant to a professor at the Sorbonne who was studying "obscure calorific rays" (electromagnetic rays beyond the red end of the visible spectrum; today we call these infrared rays). In 1880, at twenty-one, he copublished his first article.

Pierre Curie was later to write that he had vacillated over everything he had done in his life, except for his attachment to Marie Sklodowska. He was determined to make her his wife and the partner he had lost when Jacques had left home to teach at Montpelier. Marie, on the other hand, was just as determined to return to Warsaw. She told herself that she and Pierre would always be friends but she was not to be distracted from her goal. Although Marie disdained wealth and display, she tried to impress on her absent-minded suitor that one day he would have to earn a living. She knew what effort it took to lead even the most parsimonious existence. Her dedication and discipline began to galvanize this man who was drifting through life in lowly positions. For years Pierre had refused to bother completing his doctoral thesis at the Sorbonne, perhaps hampered by what Marie termed his "slow" ways. With her encouragement he presented a thesis entitled "The Magnetic Properties of Bodies at Diverse Temperatures," a superb analysis that involved working with high heat to measure tiny differences in magnetism. He found that heat had very little effect on substances that had no magnetic properties, but with those that did, there was a remarkable change of properties at a particular temperature, different for each material. Scientists still refer to this as "Curie temperature."

When Marie returned to Warsaw for the summer, Pierre suffered greatly, fearing that she might never return to Paris. He courted her by letter and knowing where her passion lay he wrote, "It would be a fine thing . . . to pass our lives near each other, hypnotized by our dreams, *your* patriotic dream, *our* humanitarian dream, and *our* scientific dream. Of all those dreams the last is, I believe, the only legitimate one." Later when Marie seemed more receptive he wrote, "If you were French you could easily become a professor in a lycée or a normal school for girls. Does this profession appeal to you?" Even later, he suggested that when she returned after the summer, they could share an apartment which would be partitioned off so that she could have privacy. She refused.

Marie wrote her sister Helena implying that Pierre's understanding of how much science meant to her touched her far more than any talk of love. When she seemed unwilling to abandon her plan to live in Poland, Pierre, though he loved France and his family, declared that he was willing to move to Poland to be with her. Pierre's mother, Claire Depouilly Curie, also pleaded his case, as did Marie's own sister, Bronya. There could be no doubt that at last she had found a man who felt her to be an invaluable part of his life. Finally, Marie acquiesced.

They were married in a simple ceremony on July 26, 1895. The reception was held in the garden of Pierre's parents' home in Sceaux, and the newlyweds left on a pair of bicycles they had purchased as a wedding present to themselves. Their "wedding tramp" lasted all summer as they toured the coast of Brittany and glided down the mountains of Auvergne. In October they returned to Paris. In that time they had become a unit. They were deeply in love.

Pierre returned to his classes at the EPCI at a salary of 6,000 francs a year. He arranged for a small space for Marie in the same building to continue working on measuring and quantifying the magnetism of various steel products. Pierre supplied and adjusted the sensitive instruments she needed, as well as imparting his superior knowledge of magnetism. The Curies already were collaborating in a way that foreshadowed what was to come. Pierre wrote, "We dreamed of living in the world quite removed from human beings," and this was a key to both their personalities. When Marie ended her relationship with Casimir Zorawski, she had written to a friend in Warsaw, "I wonder if when you see me you will judge that the years that I have just passed among *humans* have done me good or not." Her daughter Eve wrote that both her parents were made for isolation and were happiest living this "antinatural" existence. This voluntary isolation would later figure prominently in the recurrent depressive state in which Marie was to live much of her life.

CHAPTER 5

Remarkable Accidents

Scarcely four months after the Curies married, on New Year's Day 1896, a little-known professor of physics, Wilhelm Conrad Röntgen, stood before a mailbox, holding several heavy packets addressed to well-known physicists in Germany, England, France, and Austria. Röntgen was one of those nonestablishment scientists who, like Pierre Curie, had taught at a technical school, the Polytechnic at Zurich. At forty-three, after years of neglect by the scientific establishment, he was finally offered a chair in physics at the University of Munich.

Röntgen had been investigating cathode rays. In 1831 Michael Faraday had added to the three known states of matter—solid, liquid, and gas—a fourth which he called "radiant matter." He achieved this by demonstrating that if he applied an electric current to a negative terminal at one end of a glass tube from which air had been evacuated, a flow of invisible rays charged a positive terminal at the other end. But it was not until 1876 that the German physicist Eugen Goldstein coined the name "cathode rays" to describe these invisible rays, which could be detected by the light they caused in a tube. (We now know that cathode rays are negatively charged particles, or electrons.) On Friday, November 8, 1895, Röntgen as usual was alone in his laboratory. He set up an experiment in a Crookes tube using an anode and a cathode at each end (in other words, positive and negative electrodes, or conductors, by which electricity enters or leaves the tube). This was attached to a Rühmkorff induction coil (which supplied a controllable amount of electric current). When the pressure inside the tube was reduced by a hand pump to create a vacuum, and the electric current was discharged, a beam traveled between the anode and cathode and a faint glowing light could be seen. Röntgen wanted to see if any light (cathode) rays were escaping from the tube but could not tell because of the light in the room. He pulled down the shades and covered the tube with a black cardboard shield. He then repeated his experiment. Röntgen noticed something extraordinary: There was a worktable in another area of his laboratory on which rested a screen covered with barium platinocyanide, a phosphorescent material that he had used in other experiments. That faraway screen now glowed with a shimmering light.

Years later, Röntgen was asked what he thought at that moment. He replied, "I didn't think, I investigated." He repeated the experiment again with the same results. Then he moved the screen farther away and turned it away from the tube. When the electric charge was applied, the results were the same. The laboratory was dark—visible light could not be the stimulus—and the tube was shielded so that the rays could not escape unobserved. Yet rays had penetrated the shield, traveled through the air, and activated the screen. Röntgen realized that he had accidentally discovered a new type of ray. He named these mysterious rays "X-rays" (X being the symbol of an unknown quantity in mathematics).

Röntgen barred visitors from his laboratory. For the next eight weeks he worked constantly and in secret, eating and sleeping sporadically. He tried experiment after experiment. He tried to deflect the rays by placing his hand between the tube and the screen and was astonished to see a skeletal picture reflected on the screen. After that, he placed wood, tin, paper, rubber, and other materials in the same position thus revealing their interior structure in a "Röntgenogram" or "shadowgram," as it was later called. He found that plates made of glass phosphoresced more or less intensely according to how much lead they contained. (Today we know that materials with low electron density such as aluminum allow rays to be freely transmitted while those with the highest concentration of electrons, such as lead, block these rays.) Röntgen's discovery would soon find medical applications as it provided an invaluable way of looking inside the human body. Living tissue allowed the X-rays to pass through unobstructed while metal, such as a lead bullet or a swallowed pin, with more density of electrons did not.

To create permanent images, Röntgen replaced the screen with photographic plates that captured clear interior images; a closed wooden box revealed a coin inside. Finally, Röntgen asked his wife, Bertha, to come to his laboratory. He told her to put her hand on a photographic plate while he aimed X-rays at it for fifteen minutes. The bones of Bertha's hand, wearing a ring, were clearly recorded.

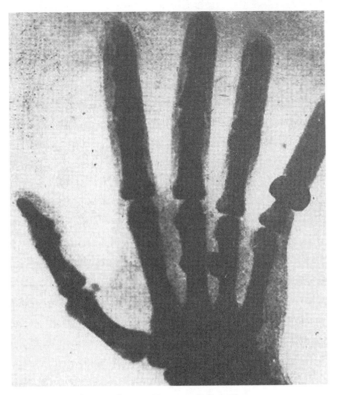

The X-ray photo of Frau Röntgen's hand.

A frightened Bertha felt a premonition of her own death. It was these shadowgrams Röntgen had mailed on New Year's Day. Two weeks later a Viennese newspaper, *Die Presse*, printed the shadowgram of Bertha's hand. It was to become one of the most famous pictures in the world.

At scientific lectures, Röntgen's medical breakthrough was applauded, but the public reaction to X-rays was one of sensational hysteria. Kaiser Wilhelm II summoned Röntgen for a circuslike demonstration of his miraculous rays, after which he was presented with the Order of the Crown. As X-rays swept the world they soon became the subject of cartoons—husbands spying on their wives by X-raying through locked doors, X-ray opera glasses that revealed naked bodies under the costumes. A New Jersey legislator moved to ban X-rays, calling their potential "lewd." A London firm sold X-ray-proof suits. One newspaper seriously suggested that medical schools use X-rays to install diagrams and formulas directly into students' brains.

Röntgen was appalled. He wrote of his disgust at not being able to recognize his own work and that his discovery was overshadowed by his newfound notoriety. He complained that fame interfered with his work. The inaugural Nobel Prize in Physics was awarded to Röntgen in 1901. Although a poor man, he gave the approximately seventy thousand gold-franc prize he received to charity. Also, he refused to patent his discovery. What happened to Röntgen was a harbinger of what the Curies would soon encounter.

In the Nobel speech Röntgen's presenter noted, "The actual constitution of this radiation of energy is still unknown." On January 20, 1896, Antoine-Henri Becquerel (called Henri), a member of the Academy of Sciences, attended a lecture on the discovery of X-rays in which the speaker hypothesized the existence of a link between X-rays and phosphorescence. Becquerel was paying scant attention, but when a phosphoroscope was mentioned he became alert. This device, which made it possible to identify substances that possessed phosphorescence, had been invented by his father, Alexandre-Edmond Becquerel.

The famous Becquerel dynasty consisted of four generations of scientists who had attended the elite École Polytechnique and had been elected to the Academy of Sciences. Both Henri's grandfather and father had been directors of the Natural History Museum (Musée d'Histoire Naturelle), and Henri had served as his father's assistant in the museum's well-equipped laboratory. In 1891, his father died. Since Henri could afford to do nothing, he remained idle for five years. If Röntgen had a great deal in common with Pierre Curie, Henri Becquerel had little. Unlike the absentminded Curie, who sometimes wore a threadbare jacket and let his mustache and beard go untrimmed, Becquerel dressed impeccably, changing his starched linen shirt twice daily. Pierre was humble and even-tempered; Henri was arrogant and irascible, with a great sense of entitlement. Pierre Curie had yet to be accredited in the scientific community, while Becquerel was born a crown prince with a silver scientific spoon in his mouth. Because of the mention of his father's invention, Henri Becquerel decided to study X-rays. Within a month he had duplicated Röntgen's experiments, producing shadowgrams of his own. He then explored substances known to phosphoresce to see if they could also produce X-rays without the need of a vacuum tube and a high-voltage electric charge.

The Natural History Museum's laboratory was well stocked with samples of various minerals and compounds. Becquerel began by placing a sample of phosphorescent uranium salts (a compound of uranyl potassium sulfate that he had prepared for his father some fifteen years previously) on a photographic plate of gelatinous silver bromide. He then put it on the windowsill to expose it to sunlight for several hours in order to "excite" the material. When he developed the photographic plate a foggy silhouette of the salts appeared. Becquerel concluded that exposure to sunlight was the catalyst that created the image. Next, along with the uranium salts he placed a copper cross on the photographic plate and prepared to expose it to sunlight. However, it rained that February day in 1896, so Becquerel wrapped the package in a black cloth and put it inside a drawer. The rain continued. By the first of March the Paris skies were still grey. Becquerel removed the package and developed the photographic plate "expecting to find the image very feeble." To his surprise, the image of the cross was clearly visible. Certainly, this result had nothing to do with Röntgen's X-rays.

Becquerel thought that these penetrating rays, which seemed to occur spontaneously, might be related to phosphorescence and have "a lifespan which could be infinitely greater" than that of X-rays. He informed the Academy of his discovery and published six papers in 1897 on "Becquerel rays." After that, believing that the subject had been "squeezed dry," he abandoned his exploration. Other scientists criticized Becquerel's work, pointing out that in 1858 Claude-Félix-Abel Niépce de Saint-Victor, a scientist who worked in Becquerel's father's laboratory, had made these same discoveries and they had come to nothing. In any case, Becquerel rays displayed neither the clarity nor the dramatic content of X-rays. Few scientists thought it worthwhile to pursue this field of inquiry.

CHAPTER 6

"The Question Was Entirely New"

Both Marie and Pierre Curie were absorbed in their work, Pierre to the extent that often he could not recall what he ate for dinner or even if he had eaten at all. His life ticked along much as it had when he lived with his parents. As was the custom of the time, women assumed all the domestic duties. Marie's garret days were over. She rented an apartment of three small rooms on the rue de la Glacière in the Latin Quarter near where she had lived as a student. This apartment, modest as it was, appealed to her because it overlooked a lush garden and light streamed in through the windows. Pierre's family supplied some meager furnishings—chairs upholstered in faded red velvet, a dining room table that doubled as a desk, a bed with heavily carved mahogany bedposts. The Curies' one luxury was fresh flowers in every room.

Marie brought to her marriage the same ardor that she had for science. She studied domestic skills as if they were scientific proposals. The student who had not known how to make soup now made gooseberry jelly and wrote down the formula including the ratio of the yield to the amount of material gathered. She purchased an expense book in which the most minute expenditures were carefully recorded. She consulted her sister Bronya constantly: How much precisely was a pinch of salt? Exactly how did one prepare a chicken for roasting? She realized that she and Pierre would need more money, so she took the necessary courses to be certified for a teaching position. At the same time, she continued her study of the magnetic properties of steel. (Two years later Marie would receive the distinguished Gegner Prize from the Academy of Sciences for this study accompanied by a much-needed award of 3,800 francs. The scientists who conferred the prize wrote that it was given for "the precision of the procedural methods of Madame Curie" and added, "Though this is an undergraduate work, it is of great interest for the creation of permanent magnets and for dynamoelectric machines.") At night, Marie attended a class on crystals in order to better understand her husband's work.

She could juggle all these responsibilities until she found herself pregnant and nauseous "all day long from morning to night." She was thirty, late in those days for bearing a first child. Ironically, though she felt miserable, she wrote that her friends thought she looked unusually healthy. On September 12, 1897, Pierre's father, Dr. Eugène Curie, delivered her of a six-pound baby girl, Irène. He noted that Marie did not cry out during the birth process but simply clenched her teeth and got on with it. Now added to an incredibly heavy schedule was a finicky baby. Marie's reaction was to start yet another journal, observing Irène's growth as if she too were a scientific project. She recorded such facts as Irène's head size, the particulars of nursing, the baby's ability to grasp an object.

Marie found herself coping with a heavy workload coupled with child care. This major problem has been ignored or glossed over in many Curie biographies. At lunchtime and in the evening Marie would rush home to nurse Irène. When she could not yield enough milk to satisfy her crying baby, though finances were strained, she was forced to hire a wet nurse, an act which gave her a feeling of failure. In order to return to work, she engaged a second nurse. Between September and December 1897, the Curies' monthly employment bill rose from 27 to 135 francs. Marie's old pattern of behavior began to assert itself—she was exhausted and depressed. Then came the panic attacks. She would suddenly bolt from the laboratory and race to the Parc Montsouris, sure that the nurse had lost her baby. On seeing Irène safe in her carriage, Marie would return to work. She was so distraught and fragmented that doctors advised she be sent to a sanatorium, but she would not leave her work, her husband, or her child. She was near a nervous collapse when luck came in the form of Pierre's father. The same month that Irène was born, Pierre's mother had died of breast cancer. Doctor Curie volunteered to come live with Pierre and Marie and look after the baby and the household. The family moved to a house on the Boulevard Kellermann on the outskirts or Paris. Soon things were running smoothly again.

Marie was now ready to begin her doctoral thesis. Like so many other scientists, she was intrigued with Röntgen's X-rays. More than 65 percent of the papers read at the Academy of Sciences in Paris were dedicated to this subject. Pierre suggested that she might instead investigate the largely abandoned Becquerel rays. In the final stages of his work, Becquerel had used Pierre's electrometer at the EPCI but seemed unable to master this delicate instrument. After Becquerel abandoned his experiments, a few other scientists had investigated these uranic rays (extremely energetic rays from uranium and other minerals), but they too found them impossible to measure. "The question was entirely new," wrote Marie.

Pierre used what influence he had at the EPCI to secure a laboratory space for his wife, but the best he could do was a small, glassed-in room on the ground floor. Marie assumed that her study of Becquerel rays would be a measuring project similar to her work on steel but more difficult. At first, Marie's results were no better than those of her predecessors. Then Pierre stepped in and worked intensely for fifteen days, modifying the electrometer, which he and Jacques had designed, to make it more sensitive to weak currents. Pierre added another of his discoveries to Marie's equipment, a piezoelectric quartz (an asymmetrical crystal that, as she noted, when compressed, "measures in absolute terms small quantities of electricity as well as electric currents of low intensity"). This addition was crucial in that Marie used it as "our device of electric conductivity." Finally, Pierre stabilized the system. Under her husband's tutelage she then spent twenty days learning how to use his equipment to measure the tiny currents generated by Becquerel rays. Without Pierre's equipment and instructions, this would have been impossible, a fact which has been largely overlooked. Even so, Pierre's equipment, state of the art for that time, was as John Joseph Thomson, director of the Cavendish Laboratory in Cambridge, remarked, "exasperating to work with." And, toward the end of his life, Lord Rayleigh wrote that all electrometers were "designed by the devil."

After thirty-five days Marie was impatient to start her experiments. She spread a thin layer of a substance containing pulverized uranium on the lower of the two metal plates, which she connected to Pierre's quadrant electrometer. This started a series of events: uranium rays bombarding the air between the plates caused the electrical changes (ionization, the release of electrons from the air molecules), which allowed a current to flow from one plate to the other. This was then transmitted through a wire to the electrometer. Inside this device, Pierre Curie had placed a thin, light blade of aluminum, known as a needle, suspended from a platinum conductive wire with a small mirror attached below. The electrical charge caused the needle to swing a tiny amount and the mirror to rotate with it. By bouncing a ray (or beam) of light on the mirror and watching it move along a graduated scale, Marie could then begin to measure extremely feeble currents.

This Rube Goldberg contraption required the scientist's complete concentration and also extreme dexterity, which

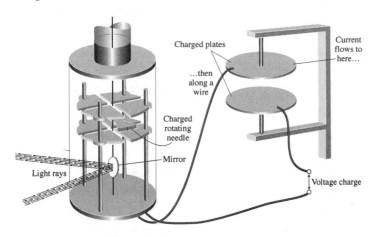

A version of a Curie electrometer used by Marie.

Marie fortunately possessed. According to her granddaughter, Hélène Langevin-Joliot, Marie was among the few scientists who could blow glass to such precise thickness that her tubes never shattered under heat and pressure.

With rare persistence and skill Marie sat, day after day, in front of her equipment. She moved only when her back hurt. Her process is even tedious to describe: Near her right hand was a piezoelectric quartz that had been stretched and weighted down by a series of small weights. After Marie had spread the test substance and charged the electrometer, she fixed her eyes on the spot of light reflected from the mirror. Then Marie lifted the weights, one by one, onto a small scale until the quantity of electric charge registering on the electrometer was identical to the opposing electric charge from the quartz. Jacques Curie noted that few people were skillful enough to do this since the operator, whose eyes were focused on the spot of light, was forced to lift the weights in an almost automatic way. "The wrist and arm must be extremely flexible." The continuous movement of the needle also required a quick reaction if one were to measure the exact amount of accumulated charge. In her left hand Marie held a chronometer (or stopwatch) which she activated to measure the intensity of the radiation in a given period of time. At last, infinitesimal amounts of currents could be measured, or "weighted."

A Curie electrometer, piezoelectric quartz, and chronometer are on display at the Curie Institute. When Hélène Langevin-Joliot was asked if she could demonstrate this intricate process she replied, "Impossible! No one at the Institute has the sleight-of-hand or the concentration to do it. In fact, I know of no one alive who has this skill."

In order to appreciate Marie Curie's skill, as opposed to that of scientists today, who enjoy worldwide collaboration and sophisticated technology, one must think of the materials and technology available to them. Experiments were conducted with wire, glue, cloth, wood, crystals, tin, glass. Heinrich Daniel Rühmkorff had made the first effective induction coil in 1851. Sir William Crookes had devised his vacuum tube in 1875. Pierre's improved electrometer was brand new. Marie and Pierre had constructed their own vacuum chamber out of discarded plywood and used a hand pump to evacuate the air.

Fortunately, a record of the Curies' work is preserved in three small dark-grey notebooks. The first notebook, partially used by Pierre Curie in 1897, contains his observations on the properties of various crystals. On January 20, 1898, Marie describes her initial efforts, which ended in failure. This is followed by notes on Pierre's adjustments as his childlike scrawl begins to mingle with Marie's small neat handwriting. Within two months, Marie was able to measure precisely the electrical activity generated by the rays emanating from various minerals picked at random. On one day alone, she tested thirteen elements, but none of them produced rays. The following week she tested several minerals that contained some uranium, as well as a sample of pure uranium. The strongest rays of all were given off by the pure uranium, so that became her standard of measurement.

Then something unexpected happened: Marie found that not only uranium and its compounds but also thorium (a mineral element discovered in 1828 by Jöns Jakob Berzelius) produced similar energetic rays. She decided she needed to expand her search and began to test many compounds, among them pitchblende, a heavy black ore mined in St. Joachimsthal, on the border of Germany and what was then Czechoslovakia from which the uranium already had been extracted, for use in creating the luminous glazes of Bohemian glass and pottery. Surprisingly, with the uranium removed the pitchblende ore still produced rays stronger than those of uranium itself. Marie remeasured. Then obsessively measured again, and again. The result was the same. To check this she tested other uranium compounds and then pure uranium powder. Once again she checked these measurements against pitchblende. The uranium compounds produced the least activity. The pure uranium was stronger. The pitchblende produced the strongest rays of all, currents four times as energetic as those of uranium alone.

Marie compared pitchblende residue with other substances: On February 24, 1898, she tested the mineral aeschynite, which contained thorium. That too was more active than thorium alone. For several days she measured the mineral calcite, which also registered a high degree of activity. Since natural calcite was hard to obtain, Marie replicated all its known elements combining uranium and copper phosphates. The radiation given off by the artificial calcite was no greater than the activity generated by the uranium it contained, while the radiation contained in the natural specimen was more than three times as active. An inevitable conclusion slowly occurred to her: There must be another element that was unaccounted for.

Once again, she verified her results, her eyes trained on the electrometer, while she manipulated the weights and timed the energetic rays with the chronometer. Often she would work late into the night, impelled forward by an insatiable curiosity. In her thoroughness, Marie was to observe, "I examined all the elements then known, either in their pure state or in compounds." She borrowed mineral samples from fellow scientists and found a rare cache of minerals in the laboratory of the Museum of Natural History. She measured not only the strength of the rays emitted but also tested to see if this energy differed in a liquid or solid state or by being exposed to light or heat. Her experiments determined that none of these conditions altered their activity.

A periodic table devised in 1869 by the Russian chemist Dmitri Ivanovich Mendeléev charted all the known elements. Mendeléev subscribed to the theory that there was nothing smaller than the atom itself, and therefore he based his table on the mass and other chemical properties of each element rather than on its atomic composition. Once a pattern was established, he could place in order the known elements while leaving spaces for the undiscovered ones that he was convinced existed. Marie Curie started her thesis in 1897, at about the same time that the English physicist J. J. Thomson discovered electrons (he called them "corpuscles" or "particles"), which he suspected were subatomic. Henry G. J. Moseley, a young scientist, discovered that the number of electrons, not mass, determines an element's atomic number. In spite of this, all his life Mendeléev stubbornly refused to accept the discovery of the electron, maintaining against all evidence to the contrary that it did not exist.

By March of 1898, Marie had established beyond doubt that several minerals gave off more energetic rays than pure uranium. In April, she wrote what was to become a seminal paper, leading to an entirely new method of discovering elements by measuring their radioactivity, thereby throwing open the door to atomic science. Neither Marie nor Pierre were permitted to present it themselves, so Gabriel Lippmann, her former professor and mentor, read it at the prestigious Academy of Sciences:

> . . . It was necessary at this point to find a new term to define this new property of matter manifested by the elements of uranium and thorium. I proposed the word radioactivity.
>
> During the course of my research, I had occasion to examine not only simple compounds, salts and oxides, but also a great number of minerals. Certain ores containing uranium and thorium proved radioactive, but their radioactivity seemed abnormal, for it was much greater than . . . I had been led to expect. This abnormality greatly surprised us. When I had assured myself that it was not due to an error in the experiment, it became necessary to find an explanation. I then made the hypothesis that the ores of uranium and thorium contain in small quantity a substance much more strongly radioactive than either uranium or thorium itself. This substance could not be one of the known elements, because these had already been examined; it must, therefore, be a new chemical element.

Marie Curie's paper contained two revolutionary observations: the assertions that radioactivity could be measured thus providing a means to discover new elements, and that radioactivity was "an atomic property." At that time most scientists, including the Curies, still believed in Democritus' fifth-century B.C. definition of the atom as "the smallest part of matter" which could not be further decomposed. Marie did not use the term *atomic property* as we do today, but meant simply that radioactivity was a property of the atom itself. Dr. Spencer Weart, director of the Center for History of Physics in Maryland, provides a clear explanation:

> The properties of a metal, for example, include its silvery shine, its brittleness, its heat conductivity and capacity. (It feels cold when you touch it, etc.) None of these are the properties of a single atom such as iron. Realizing that radioactivity could not be changed by any chemical procedure—dissolving in acid or water, heating or cooling, etc.—and therefore it was an atomic property was Marie Curie's important intellectual contribution to science. (Of course, we now realize radioactivity is a property of the nucleus, not the entire atom.)

Ernest Rutherford, who would contribute greatly to the exploration of radioactivity and the atom, explained, "I was brought up to look at the atom as a nice hard fellow, red or grey in color according to taste." And Lise Meitner, a chief physicist at the Kaiser Wilhelm Institute for Scientific Research and a discoverer of atomic fission, remembered that she had been taught that the atom was made up of "solid unsplittable little lumps."

The word *atom* itself means "indivisible," or more technically derives from the Greek words for *not* and *to cut*. As long ago as 98 B.C., the notable Roman poet Lucretius wrote that atoms were "the only eternal and immutable entities of which our physical world is made." Even then, there was philosophical if not scientific debate. In 322 B.C., Epicurus wrote that atoms could not be divided. However, Anaxagoras and Aristotle postulated that all matter was divisible and continuous, and later Descartes wrote that even if man did not have the power to divide the atom, God did, so therefore it *must* be divisible.

A century before the Curies began their work, the British chemist John Dalton had set down the rules to which scientists adhered. Atoms, he stated, were composed of tiny indivisible particles of matter and were the smallest unit of an element. An atomic property, Dalton said, was associated with a single atom of an element. By 1873, Dalton's view had become inviolate. The august James Clerk Maxwell, who formulated the laws of electromagnetic forces, wrote, "Though ancient systems may be dissolved and new systems evolved out of their ruins, the molecules [atoms] out of which these systems are built . . . [are] the foundation stones of the material universe and remain unbroken and unworn. They continue to this day, as they were created."

It was only in 1876 that two professors at the Sorbonne, Charles Wirtz and Charles Friedel, Jacques Curie's mentor, introduced the study of the atom into the science curriculum. The model of the atom they used was a solid lead ball. The caption read, "The smallest molecule of matter." The classes were sparsely attended. In short, before Marie's observation, it was thought, as Rutherford put it, that the subject of the atom was "bankrupt." In any case, Marie Curie's discoveries were met with indifference. Who was this person? She was a scientist manqué who had not yet completed her doctoral thesis. She was a Polish émigré who had worked as a governess. She was married to an industrial teacher. She was a woman.

CHAPTER 7

"The Best Sprinters"

As Manya Sklodowska shyly gave her father her high school medal for the best pupil of 1883, a boy of eleven, Ernest Rutherford, stood on the porch of a New Zealand farmhouse while a thunderstorm approached. His father, awakened by the storm, went downstairs to join his son. What was he doing? Ernest replied that he had figured out that by counting the seconds between the lightning flash and the thunderclap, and allowing one second for the sound to travel 400 yards, he could tell how close they were to the storm's center. Until then Ernest, one of twelve children of a potato farmer, had like Pierre Curie been considered slow. Home-schooled, at eleven he could read but not write. At twelve, he was lucky enough to find the first of a series of gifted teachers who inspired him to learn. When he received his first full scholarship, he told his mother, "I've dug my last potato."

In 1898, as Marie Curie concentrated on her thesis, Rutherford arrived at the Cavendish Laboratory of Cambridge University. In New Zealand, chemistry and physics were so little regarded that he had been one of only three students at Nelson College to attend these courses, but he was determined to make science his life. At twenty, Rutherford had become interested in the discoveries of the German scientist Heinrich Rudolf Hertz, who had demonstrated that electric waves could be sent through space, as well as the work of Guglielmo Marconi, who had harnessed these waves to produce a radio signaling system though no efficient receivers as yet existed. Rutherford thought that if a device could be magnetized and demagnetized by alternating the current, it would be sensitive to these waves and thus could be used as a superior receiver. In an early experiment, he passed an alternating current through the hollow center of a copper wire coil into which he had inserted a series of sewing needles. This simple device proved to be highly efficient in receiving these waves. Rutherford's experiments, like those of Marie Curie, were influenced by an original and commonsense clarity of vision.

When J. J. Thomson heard of Rutherford's work, he offered him a position. Rutherford borrowed the money for steerage passage to England, where he hoped to sell his radio device to earn enough to marry his fiancée, Mary Newton. Thomson told Rutherford such crass considerations as money were "without honor" and not worthy of a scientist. (Mary Newton would wait five more years until Rutherford felt financially secure enough to marry her.) Thomson had manipulated Rutherford for reasons of his own. The previous year, Crookes had observed that when a high-voltage electric current was passed through the vacuum tube he had devised, it engendered a discharge of cathode rays, but no scientist had been able to determine their exact composition. Thomson postulated that these rays might be emanating from minute pieces of atoms. He conducted three experiments: The first established, through the use of a magnet, that the electric charge in the rays could not be separated from the rays themselves. In the second, he was able to bend the cathode rays by creating an electric field within a vacuum tube. In the third, he measured how much energy the rays carried, and calculated the ratio of the mass to the electric charge.

From these experiments Thomson determined that cathode rays had great momentum, which meant they possessed mass (since it was known that momentum equals mass times velocity). He concluded that these rays were composed of negatively charged fragments of atoms, projected with a velocity so great that their measurable mass was 1/1,800 of a hydrogen atom, which was the lightest-known atom. Although Thomson had discovered the first subatomic component, he was to write, "At first there were very few who believed in the existence of these bodies smaller than atoms."

Under Thomson's influence, Rutherford began to study X-rays and the radioactive Becquerel rays. In his own landmark experiment, he placed thin aluminum foil (five micrometers thick) between a uranium compound and a charged plate and discovered that two different types of radiation were emitted. These he called alpha and beta rays in the sequence of their ability to penetrate the foils. The weaker, alpha rays (which are positively charged particles containing two protons and two neutrons—the nucleus of a helium atom) could not even penetrate a thin layer of cardboard. Beta rays (which proved to be Thomson's corpuscles and came to be called electrons), had a greater ability to pass through matter. Rutherford also observed a third even more powerful ray. In 1904, Paul Villard named this a gamma ray to follow Rutherford's system. Both gamma rays and X-rays proved to be electromagnetic and had the ability to penetrate many material barriers, but the radioactivity of gamma rays was of a much higher energy.

The race to discover new elements that produced more radioactivity than uranium alone had begun. In his straightforward way, Rutherford wrote his mother, "I have to publish my present work as rapidly as possible in order to keep in the race. The best sprinters in this road of investigation are Becquerel and the Curies in Paris, who have done a great deal of very important work on the subject of radioactive bodies."

Although one may think of scientists as abstracted from worldly acclaim and rewards, in fact that is rarely the case. Pierre Curie was an exception. He was indifferent to how quickly their research progressed, or even who took credit for it. Marie was not so sanguine. As if the Furies were at her back, she wrote, "I had a passionate desire to verify [my] hypothesis as rapidly as possible." She was distressed to learn that the Italian scientist Emilio Villari had measured the radioactivity of uranium and like Marie herself was using it as a standard in his investigation of other substances. This was followed by the news that Gerhard Carl Schmidt had beaten the Curies by three weeks in publishing his findings on the radioactivity of thorium.

To forge ahead, the Curies needed more help and more money. Charles Friedel, who knew Pierre well, nominated him for a professorship in mineralogy at the Sorbonne, which would yield 12,000 francs, double his present annual salary, as well as access to a well-equipped laboratory that both Curies could use. Although Pierre was by far the most knowledgeable person in Europe on the subject of crystals and piezoelectricity, he was rejected by the Sorbonne in favor of his friend Jean Perrin, who readily admitted that Pierre was the superior scientist. But Perrin was a member of the establishment and had attended the École Normale Supérieure, while Pierre had taught at an industrial school. Pierre's loss was Marie's gain. Within two weeks of this rejection she wrote, "He abandoned his work on crystals to join me in the search for this unknown substance."

Of the materials she tested that were more radioactive than pure uranium or thorium, pitchblende ore (whose radioactivity had measured four times that of uranium) seemed the easiest with which to work. Marie separated its various elements "by the ordinary means of chemical analysis," and with Pierre's sensitive equipment measured the radioactivity of each separate product. Then the Curies distilled the most radioactive elements further. With each new distillation the radiation became stronger.

As the separation process progressed, the composition of the pitchblende began to reveal itself. "We soon recognized that the radioactivity was concentrated principally in two different chemical fractions." One behaved chemically like bismuth, the other like barium. Although the Curies could not afford to hire helpers, the excitement of what was going on in Marie's makeshift laboratory attracted other scientists. When Marie was having trouble eliminating the lighter elements, the EPCI laboratory director, Gustave Bémont, a master chemist, suggested that she boil each distillate and then cool it. In this manner, the solution formed crystals and the lighter elements crystallized first. She could then separate the elements according to their tendency to form crystals at different temperatures. This technique Marie called "fractionation." Soon Bémont was spending his spare minutes collaborating with the Curies. André Debierne, Pierre's former student, who held a position as a *préparateur* (assistant) in one of the Sorbonne's state-of-the-art laboratories, arrived almost every day after work to help prepare the chemicals used in Marie's treatments. Debierne, who proved to be a gifted chemist, was to work with Marie Curie for the next forty years. A shy, unassuming man, he was to become her loyal right hand, always there for her professionally and privately.

As news of the Curies' experiments began to circulate throughout in the scientific community, Henri Becquerel's interest was rekindled. Here was a way for him to demonstrate that he was not just a member of a dynasty of establishment scientists but could accomplish great work on his own. Once again he left his own fine laboratory at the Museum of Natural History to study strong distillates, borrowed from Marie, with Pierre's equipment at the EPCI. Becquerel amplified on Marie's findings that radioactive substances remained the same in a liquid or solid state. Because of his father's experiments he knew that phosphorescence remained even after the activating stimulus was withdrawn. Therefore, he concluded that radioactivity must be a form of phosphorescence. For a brief period, Marie thought that radioactivity could be a form of "disintegration in the atom," but when Pierre found this hypothesis ridiculous and agreed with Becquerel, she abandoned it and joined Pierre in his search for outside forces.

The Curies' success was now bonded to that of Becquerel. Henri used his influence to secure two successive Academy of Sciences grants for Marie but did not see fit to inform her directly. Instead, he wrote to Pierre Curie expressing his "sincere congratulations" and requested that he "inform [his] wife and present my respectful compliments." Pierre complained to a friend that Becquerel looked down on Marie because she was a woman. Although beholden to Becquerel, Pierre disliked and distrusted him and felt Henri only pretended to be their friend for his own self-aggrandizement. Pierre wrote to his friend the scientist Georges Gouy that he and Marie were "fed up" with this imperious man. Three of Pierre's friends encouraged him to apply for membership in the Academy of Sciences, which would allow him to request grants and read his own papers. Against his nature, Pierre called on influential scientists to gain their backing, though he complained that this took too much of his time away from work. When he was rejected, he blamed Becquerel. "Even though declaring for me, I am convinced he [Becquerel] played a double game. I am sure he was delighted that I didn't get in," he wrote to Gouy.

The Curies' work progressed at a rapid pace: Pierre scrawled in their workbook that Marie had produced a substance accompanying bismuth that was 17 times more radioactive than pure uranium alone, then two weeks later 150 times as radioactive, then 300, then 330. The radioactivity of this last substance was so great that Marie was convinced she had discovered a new element. But how to confirm it? A sure way was by a method known as spectroscopy and the EPCI was fortunate in having a resident expert in this field, Eugène Demarçay. Spectroscopy involved the heating of an element until it became a glowing gas and then refracting the light it emitted through a prism. This resulted in rainbow patterns of light, or spectra. No two elements produced the same pattern of light. Eight new elements had been found by this method. Demarçay tested Marie's substance but said it was not sufficiently pure to produce a spectrum. Though bitterly disappointed, she marched back to the laboratory. Within ten days she had, in her own words, "obtained a substance 400 times as active as uranium alone." Demarçay tested this substance, but once again could not produce a clear spectral line.

Marie felt she could wait no longer. Her work was usually

exhaustive and irrefutable, but this time her scientific caution was overridden by her determination to win the race to discover a new element. She had blazed a path with her method of inquiry, only to find that German, Italian, and English scientists were employing it to try to find new elements before she could do so. Having established a measurable amount of radioactivity, she felt she had sufficient proof of the element's existence. Marie promptly announced her discovery and named it polonium for her beloved country. In July, the Curies wrote a third paper about polonium, cautiously noting that Demarçay had not as yet found a clear spectral line for this element, nor had it been separated from bismuth. Pierre and Marie tactfully included their benefactor, writing, "If the existence of a new element is confirmed, this discovery will be due entirely to the method of investigation provided us by Becquerel rays."

Marie and Pierre posing in their laboratory.

One element had been found, but amid growing excitement Marie set out to find a second element, the one that behaved almost exactly like barium. It was difficult to remove the barium, but four months later on a December morning she finally produced a substance whose radioactivity registered 900 times greater than that of pure uranium. Marie feared that this immense power might quickly disintegrate, so she threw a sweater over her shoulders, raced up the stairs, and burst into Demarçay's laboratory. This time he found a clear, unique spectral line. On December 19, 1898, a notation in the Curies' laboratory notebook reads, "Radium" (a name derived from the Latin *radius*, meaning "ray").

Six days later a final paper was presented to the Academy by Becquerel and published in the scientific journal *Comptes rendus de l'Académie des sciences*. Written by the Curies and the EPCI chemist, Gustave Bémont, it announced the discovery of radium, "a new, strongly radioactive substance contained in pitchblende." Their finding was endorsed by Demarçay's report of the spectral line "known to no other substance." In the dash to glory the entire process of discovering both polonium and radium had taken only one year.

In the future, Marie Curie's discovery of polonium would be lost in time. Radium would become her "colossal achievement." But, in fact, her greatest achievement was in employing an entirely new method to discover elements by measuring their radioactivity. In the next decade scientists who located the source and composition of radioactivity made more discoveries concerning the atom and its structure than in all the centuries that had gone before. As the astute scientist Frederick Soddy said, "Pierre Curie's greatest discovery was Marie Sklodowska. Her greatest discovery was . . . radioactivity."

CHAPTER 8

"A Beautiful Color"

Polonium and radium had been discovered, but only
theoretically. Physicists were somewhat prepared to
accept the Curies' discoveries because they were working with the properties of rays, but chemists were not. Until
there was an actual substance that could be seen, handled,
and weighed, they would remain skeptical. Marie wrote,
"There can be no doubt of the existence of these new elements
but to make chemists admit their existence, it was necessary
to isolate them." At the time she wrote this, she had no idea of
the magnitude of the task that lay before her, one that would
require limitless vision, skill, persistence, and the dedication
of a zealot.

On the December day that she had named radium, Marie
had written, "The new radioactive substance certainly contains a strong portion of barium. In spite of that, the radioactivity is considerable. The radioactivity therefore must be
enormous." Marie calculated that radium, present only "in
traces," must be "several hundred times more active than ura-

nium." She was mistaken. One tenth of a gram of pure radium chloride would prove to be ten million times more radioactive than pure uranium. Four years later, Pierre Curie said that had it been his choice, he never would have attempted the task of isolating radium. For Marie there was no choice.

Marie's quest to isolate pure radium has taken on mythic proportions. Madame Curie presumably suffered years of backbreaking toil without money, without assistance, enduring the scorn of fellow scientists, to isolate pure radium salts and apply her discovery to the cure of cancer. Over the years this legend has been burnished, not in the least by Marie herself when, years later, she sought money to continue her research. The Curie papers, diaries, letters, and related documents reveal an equally heroic but more complex and nuanced story.

In this journey of discovery, Marie and Pierre were equally involved as they were in life. They thought that their task would be easier if they worked on different aspects of the problem. Pierre took over the physics, exploring the origin and nature of radium's activity. Marie acted essentially as a chemist who, as her daughter Irène was to observe, "had the stubborn desire to see salts of pure radium, and to measure radium's weight."

In the first few months Marie chose to work with small quantities of pitchblende ore residue:

I extracted from the mineral the radium-bearing barium and this, in the state of chloride, I submitted to a fractional crystallization. The radium accumulated in the least soluble parts. . . . By the end of the year, results indicated clearly that it would be easier to separate radium than polonium; that is why we concentrated our efforts in this direction.

It was becoming evident that to produce any measurable quantity of radium, larger quantities of pitchblende residue would be required. On the basis of their discovery of radioactivity, radium, and polonium the Curies applied to the Sorbonne for space to work in one of their numerous buildings, a request which usually was freely granted to scientists. They were refused. The director of the EPCI admired the Curies and was sympathetic to them but had only limited facilities. He did the best he could, offering them the use of a cavernous abandoned hangar across a wide courtyard from their present storeroom laboratory. The hangar had been used by students for human dissection, but it had become so leaky and dusty that the corpses were moved to another location. When Ernest Rutherford saw these quarters four years later he remarked, "You know it must be dreadful not to have a laboratory to play around in," and the Nobel Prize–winning chemist Wilhelm Ostwald noted, "I insisted on seeing the laboratory. . . . It looked like a stable or potato cellar and if I had not seen the worktable with the chemistry equipment I would have thought it was a hoax." But in a lucky unforeseen circumstance, without the open courtyard the tons of pitchblende residue that eventually were needed for the isolation process never could have been accommodated.

There was no money to buy the needed ore but Pierre went to Eduard Suess, president of the Academy of Sciences of Vienna, and asked what had happened to the pitchblende ore residue from which the uranium had already been removed. Suess found out that the residue had not been destroyed but had been abandoned, a mountain of sludge deep in the forest of St. Joachimsthal. With ingenuity, Pierre persuaded Suess to arrange for the Austrian government to give them this seemingly worthless material at no cost. Then he went to the Baron Edmond de Rothschild and asked him for a donation to cover the cost of transportation. Over the next four years, Rothschild, who at that time remained anonymous, gave repeated donations for this purpose.

When the first large shipment of pitchblende residue was dumped in the yard outside the hangar, Marie scooped up a handful of "the brown dust mixed with pine needles" and held it up to her face. Then, shaking with emotion, she rushed inside to begin work. The initial steps in processing the residue involved physical strength more fit for a factory worker than the frail Marie. She spent weeks stirring the boiling residue to form the first reductions. Only then could the chemical washings begin and after that the splitting process and measurements of the distillates. It was becoming evident that an infinitesimal amount of radium sent off such strong rays that to isolate the substance itself might require tons of material. After six months of watching his wife labor, Pierre became alarmed at her obsessive activity and increasingly fragile condition. Although they were still pressed for money, in June of 1899, Pierre nevertheless hired André Debierne full-time to assist Marie. It was becoming increasingly evident to Pierre that the goal his wife had set for herself would be impossible to achieve. But Marie would not give up. It is little known that Pierre, understanding his wife's fixation, turned this impossible task into one that was only brutally difficult.

Marie had written that "our research on new radioactive substances gave birth to a scientific movement." Scientists in Germany, Canada, England, and Australia were now anxious to buy strong radioactive materials. Pierre cleverly convinced the owners of the Central Society for Chemical Products in France to pay Debierne's salary as well as that of several other workers the Curies needed to hire. In exchange, they granted the Central Society a portion of their strong distillates to sell to these eager scientists.

With the help of the Central Society, the method of extraction escalated to an industrial scale. André Debierne was put in charge of this operation. Under his capable direction, within three and a half months the Central Society's factory had treated a ton of pitchblende residue, subjecting it to innumerable washes with acid, alkaline salts, and water. Each ton thus processed required fifty tons of rinsing water to achieve a bromide that was fifty times as radioactive as that of uranium. Then Marie took over, starting each time with twenty kilograms of material from the Central Society. With Pierre's help she performed the fractionations and measurements, thus producing stronger and stronger amounts of radium.

In spite of the repetitious, difficult labor, Marie felt that this, the "heroic period" of their exploration, was exhilarating:

We were very happy in spite of the difficult conditions under which we worked. We passed our days at the laboratory, often eating a simple student's lunch there. A great tranquility reigned in our poor shabby hangar; occasionally, while observing an operation, we would walk up and down talking of our work, present and future. When we were cold, a cup of hot tea, sipped beside the stove, cheered us. We lived in a preoccupation as complete as that of a dream.

Yes, these two were dreamers. Frequently at night, Pierre and Marie would walk hand in hand the five blocks back to the laboratory, drawn by the mysterious element that these two scientists viewed with romanticism. "I wonder what *It* will look like?" Marie asked. Pierre answered, "I should like it to have a beautiful color."

The romance with radium was to heighten their partnership. As their astute daughter Eve was to observe,

> The days of work became months and years: Pierre and Marie were not discouraged. The material which resisted them, which defended its secrets, fascinated them. United by their tenderness, united by their intellectual passions, they had, in a wooden shack, the anti-natural existence for which they had both been made.

The marriage Marie had reluctantly entered had turned into an intense but calm love affair. Marie felt that without speaking she and Pierre understood each other's thoughts. She wrote her sister Bronya that she was married to the most wonderful man in the world. As Pierre passed her in the laboratory he would stroke her hair. At night they slept in each other's arms. He comforted her on a terrible day when from nervous exhaustion she spilled three months worth of her precious distillates on the floor. It was he who calmed her when she wrongheadedly argued with Demarçay about radium's spectral line, saying gently, "Oh come now, Marie." In his vest pocket Pierre carried his favorite photograph of her, which had been taken shortly after their meeting. Marie seems lost in thought, her curly blond hair tamed into a chignon; she wears a striped blouse and a skirt that emphasizes her tiny waistline. Pierre called this photograph "the good little student."

At home, seemingly there were few problems. Irène had formed a close bond with her grandfather Eugène Curie. When she was three and a half, she asked him why her mother

"The good little student."

had to leave her to work all the time since the mothers of her friends stayed at home. Dr. Curie tried to explain that she was doing great work, and took Irène to her parents' laboratory to show her what they did. The child was dismayed, calling their shabby laboratory "that sad, sad place." Irène, like many children with remote mothers, demanded Marie's attention. She would cry when Marie left the room and cling to her skirts. She refused to go to sleep without her mother's kiss.

The conversation in the Curie household centered on science and how to support their research. The rent on their house on the Boulevard Kellermann was 4,500 francs a year. Pierre agreed to teach yet another class for 2,000 francs, and Marie became a physics professor at the École Normale Supérieure des Jeunes Filles at Sèvres, an elite academy that trained brilliant young women who might then go on to be teachers, scientists, and so forth. She was the only woman on the faculty. The commute to Sèvres was an hour and a half each way, taking precious time from her laboratory work, but they needed the money.

The search for pure radium was now entering its third year. Eight tons of pitchblende residue had been processed along with four hundred tons of rinsing waters and thousands of chemical treatments and distillations. According to Marie, these substances were "arranged on tables and boards. . . . At night . . . from all sides we could see slightly luminous silhouettes, and these gleamings, which seemed suspended in darkness, stirred us with new emotion and enchantment. . . . It was really a lovely sight . . . like faint fairy light." The light was caused by radioactive atoms releasing their energy, but the Curies were unaware that exposure to these substances was affecting their health. A century later the Curies' personal items, clothes, and papers were still radioactive. Pierre was now suffering from rheumatism, which he attributed to the dampness and leaking in the hangar. Marie was rapidly losing weight. Their friend, the scientist Georges Sagnac, wrote to Pierre:

> You hardly eat at all, either of you. More than once I have seen Mme. Curie nibble two slices of sausage and swallow a cup of tea with it. . . . Her own indifference or stubbornness will be no excuse for you. . . . She is behaving at the present time like a child. . . . It is necessary not to mix scientific preoccupations continually into every instant of your life as you are doing. . . . You must not read or talk physics while you eat.

This warning was ignored.

At this point however, had it not been for Pierre's indecisive nature, France would have lost its most precious scientists. In the spring of 1900, the University of Geneva offered Pierre a position as a physics professor with a well-equipped laboratory and a salary of 12,000 francs a year. Marie was offered a lesser but substantial position and salary. Earlier Pierre had refused the Legion of Honor, saying, "I ask you most kindly, to thank the Minister and inform him that I do not need to be honored, but rather I most certainly need a laboratory." The laboratory that was unobtainable in France was freely offered in Switzerland, but it would mean leaving France. Now began an agonized minuet, reflecting Pierre's nature as well as the Curies' desperation.

Pierre met the dean of the university and expressed a positive interest in the position. Then he hesitated. Two months later, he refused the position. Then he reconsidered and accepted. They traveled to Geneva. They loved the location, the mountains, the clear air, and most of all the promised laboratory. On returning to Paris, Pierre changed his mind. He refused again. Then a few weeks later, after an exhausting day, he wrote the dean accepting "definitively" and resigned his position at the EPCI. Just as the Curies were about to leave France, two scientist friends intervened. At long last, Pierre was offered a position as a physics instructor and lecturer at the Sorbonne annex on the rue Cuvier. He accepted the position.

Marie returned to her tedious work of seeking to isolate the elusive radium. As her distillates became stronger and stronger, so did her determination to produce pure radium salts and measure radium's mass. Pierre reminded her that summer had come and it was time for their family vacation. He said that Irène had seen little of her parents and that somehow both of them were ill. He was suffering from unexplained bone pain, and Marie's doctor suspected that she might have contracted tuberculosis and had prescribed rest and country air. Marie would not budge.

In July 1902, as Marie entered her fourth year in their makeshift laboratory, she made yet one more fractionation and measured its radioactivity. Finally, she had produced a specimen that contained too little barium to influence radium's weight. In the numbing years of measuring and remeasuring, of thousands of fractionations, of ten tons of processed pitchblende residue, she had produced an amount of pure radium so small that it resembled a few grains of sand.

Many of us have been led to think of radium in terms of beakers of luminous material. In a well-known photograph, Marie holds such a beaker in her left hand, another in her right, and yet another is placed on a table beside her. Except for its radioactivity, radium has the same chemistry as calcium. Like calcium it will form chlorides, or sulfates, or carbonates. These beakers do not contain pure radium, but the interim fractionations. (Luminous paint, for example, requires only a ratio of one part pure radium chloride to approximately thirty-five thousand parts of zinc sulfide.) Several biographies and scientific books assert that the pure radium she isolated would fill one-quarter to one-half a teaspoon, a tiny amount, but in fact the amount was even smaller: A minute *one-fiftieth* of a teaspoon. Her granddaughter Hélène Langevin-Joliot explains:

> Teaspoons may have very different volumes. Assuming the length is about 25 mm, the width 15 mm, and the mean depth 3 mm, that would give slightly more than one cubic centimeter. Marie used 1 decigram of radium chloride (density 4.9 g per cm^3) to determine radium's atomic weight. Therefore the 1/50 of a teaspoon that she produced is much less than quoted. [Marie] prepared . . . in the following years, up to 1 gram, which would be closer to a quarter of a teaspoon. . . .

This minuscule amount, however, was so powerful that, as Pierre's experiments were to prove, the energy radium released produced enough heat to bring its weight in water from freezing to boiling in only one hour. Using Einstein's equation of E (energy) = mc^2, the E (energy) of only a pound of such mass (m) multiplied by 448,900,000,000,000,000 (the speed of light (c, 186,282 miles per second) squared (^{2}) yields 10 billion kilowatt hours of energy so intense as to equal the annual output of a vast power station or, if released all at once, a medium-scale atomic bomb. In *Comptes rendus*, Marie Curie announced in dry academic terms that she had isolated radium. She placed it correctly on Mendeléev's chart as number 88.

In Warsaw, Wladyslaw Sklodowski was dying. Two months before her public announcement she wrote him of her triumphant discovery. Earlier she had written, "My father, who in his own youth had wished to do scientific work, was consoled . . . by the progressive success of my work." But was he? Was anything enough for this man's ambition? Marie's father answered, "You are now in possession of pure radium salts. If we consider the amount of work done in obtaining this, it would certainly be the most expensive of chemical elements. What a pity it is that this work has only theoretical interest." The father who had driven his children to fulfill his broken dreams died six days later, never to see how wrong this pronouncement would prove to be.

CHAPTER 9

"What Is the Source of the Energy?"

The Paris Exposition Universelle of 1900 was stupendous. When President Émile-François Loubet invited the mayors of France for a banquet, 20,777 showed up and occupied 606 tables. Forty countries were represented and 210 pavilions were built on 277 acres covering a quarter of Paris itself. The French had outdone the 1893 Chicago World's Fair by attracting the Olympic Games. The Eiffel Tower, built for the Exposition of 1889, loomed as a beacon to lure fifty million visitors. At dusk, with a flick of a switch, the Grand Palace of Electricity was bathed in the light of 5,700 incandescent bulbs. Electricity powered a train that ran around the perimeter of the exposition; a *trottoir roulant* (moving walkway) two miles long allowed visitors to glide past the pavilions. The Grand and Petit Palais, both built with flush toilets and electric lights, were to become permanent installations. At hundreds of stands, postcards made by an innovative process, an acid bath followed by an etch-powdering process, produced images of great clarity. A new century had arrived and a wondrous world was unfolding, revealing the energy of the unseen.

The phenomenon of electricity attracted millions of visitors, but for the cognoscenti it was radioactivity that claimed worldwide interest. As part of the celebration of science the Curies and Becquerel were asked to consolidate their findings and present them on the afternoon of August 8. Marie was pleased with the prospect of exposing their work to such a vast audience. Pierre was indifferent but did most of the writing. Becquerel presented the history of his discovery of uranic rays and his further experiments on the conduction of radioactivity through air. The Curies' paper, "The New Radioactive Substances and the Rays They Emit," observed, "Spontaneous radiation is an enigma, a deeply astonishing subject. . . . What is the source of energy from Becquerel's rays? . . . It remains undetectable. Must we look for the energy source in radioactive bodies themselves or should we look outside of them?" This question was pursued by many scientists, but especially Ernest Rutherford.

Having left the Cavendish Laboratory for McGill University in Montreal in 1898, Rutherford set out to study the radioactive gases (which he named "emanations") from uranium and thorium salts. He was disheartened to find that only two weeks before his own paper was published, Marie and Pierre Curie had published similar findings. In their paper "On Induced Radioactivity and the Gas Activated by Radium," they observed that "radioactivity gradually transmits itself through open air from the radiant material to active bodies." This radioactivity had spread to everything in their laboratory—tables, chairs, paper, equipment, clothes. All became contaminated, which created what Marie called a "deplorable state. . . . In our laboratory the situation has become acute and we no longer have any apparatus properly insulated."

Pierre, who took the position that this radioactive transmission did not emanate from inside the atom but took place in the surrounding environment, suggested that in all probability "a radioactive atom is a mechanism that in every instance, has the capacity to release energy outside itself." Pierre was also convinced that "radiation does not diminish with time." Becquerel added a note that this transfer of radioactivity was "a kind of phosphorescence."

Ernest Rutherford would have none of it. In his mind the Curies were wrong. The scientific dueling between the Curies and Rutherford would last more than three years. Marie began a series of experiments which seemed to prove that "no change occurs in this material which radiates the energy." (At the time, this seemed a reasonable assumption since these transformations took place so slowly that they were imperceptible.) In January 1902, she and Pierre co-wrote a paper that contained a thinly veiled attack on Rutherford (without mentioning his name) for being "premature" in his belief that radioactivity came from within the atom. Rutherford, they felt, was dashing ahead recklessly, playing the hare to the Curies' tortoise. They wrote,

As for the origin of radioactive energy, we could come up with . . . two very general hypotheses: 1) Each radioactive atom contains, in the potential energy state, the energy that it releases. 2) A radioactive atom is a mechanism that, in every instance, has the capacity to free the energy outside itself. For the first hypothesis [we find] negative results. . . .

For the second, the study of unknown phenomena, we hypothesize very generally and progress little by little depending on the experiment. This methodical and precise path is necessarily slow. Or, on the contrary, we could make a daring hypothesis to confirm a margin of error, next to a degree of truth.

Marie Curie and Ernest Rutherford often replicated each other's experiments, but came to different conclusions. Rutherford requested a sample of the Curies' strong distillates of radium and thorium. Although these samples were sold by the Central Society at a high price, and the Curies were competing scientifically, out of professional courtesy they obliged him. Rutherford enlisted Frederick Soddy, a junior member of the McGill chemistry department, to help study thorium X, an isotope of radium. (An isotope is chemically identical to a given element because it has the same number of protons, but it is different in its mass because it contains a differing number of neutrons.) They observed that the radioactivity of thorium X decreased by half in four days. In this process of "atomic disintegration" they noted four successive transformations followed by "subatomic chemical formations within the atom itself." The following year they proved that heavier elements belonging to the same family (such as uranium, thorium, polonium, and radium) were unstable and continuously transmuted themselves into lesser radioactive elements until the final product became lead. The time it took for half the quantity of a radioactive substance to disintegrate or decay into another substance constitutes what is called its half-life, which can be minutes, hours, days, or years. Judging from the energy released, they calculated that the half-life of radium was "approximately two thousand years."

Against the resistance of the Curies, Rutherford and Soddy had begun to probe the interior of the atom. So startling was their work that during the seminal experiment on the half-life of radioactive materials, the young Soddy turned to Rutherford and exclaimed with awe, "This is transmutation: The thorium is disintegrating and transmuting into argon gas." To which Rutherford replied, "For Mike's sake, Soddy, don't call it *transmutation*. They'll have our heads off as alchemists."

In spite of evidence to the contrary, Pierre Curie clung to his "outside" theory and would not acknowledge the transmutation process. Once again he published his beliefs. This so disturbed Rutherford that he wrote, not in his native English but in French, to the editor of the *Philosophical Magazine*, refuting every one of Pierre Curie's arguments and concluding, "M. Curie apparently has not seen [my] last article. . . . In light of these results . . . the alternative theory proposed by Monsieur Pierre Curie . . . seems useless to me."

In 1904, when Pierre Curie (in collaboration with an assistant, Jacques Danne) finally replicated the experiments of Rutherford and Soddy, he had no choice but to reluctantly accept their conclusions. In March of that year, Pierre published a paper "adopting Mr. Rutherford's manner of seeing." Marie, who understood her husband as no other, realized that this humble physicist, who disdained worldly goods and fame, nevertheless possessed an Achilles heel. She wrote, "Every good subject brings him pleasure, but in this domain he expects to have priority."

Radioactivity, however, had implications far beyond the

question of its origins. Rutherford and Soddy observed the immense energy emanating from the discharge of subatomic particles and gamma rays within the atom. Soddy calculated that the energy released during "radioactive change must . . . be at least twenty-thousand times, and may be a million times, as great as the energy of any molecular change."

In June of 1903, Marie defended her doctoral thesis before a scientific board, the first woman in France to achieve this level. Bronya came from Poland, took one look at her sister, and whisked her off to a dressmaker. Marie chose a black dress that would not show stains in the laboratory. A small group of scientists and friends attended the event, including a proud Pierre; Marie's mentor, Gabriel Lippmann; physicists Jean Perrin and Paul Langevin; as well as her father-in-law, Dr. Eugène Curie, and women from the classes she taught at Sèvres.

By chance Ernest Rutherford was in Paris that day on his long-delayed honeymoon with Mary Newton. They joined the Curies in a celebratory dinner. In spite of their scientific differences, or perhaps because of them, Rutherford liked Madame Curie. He admired her technique, which was like his own in that it eliminated anything superfluous to the question, as well as her mind, which was willing to question. After the last toast, the group strolled out into the garden. In the dark of the night, Pierre reached in his vest pocket and drew forth a glass tube of radium bromide. Its magnificent luminosity gleamed as he held it up, illuminating an expression of rapture on Marie's face. Rutherford observed that it also illuminated the cracked flesh and burned skin of Pierre's irrevocably destroyed fingers.

CHAPTER 10

"I Will Make Him
an Help Meet for Him"

On December 10, 1896, the industrialist Alfred Nobel died, leaving his fortune (in 1867 he had patented dynamite) to be administered by the Swedish Academy to distribute prizes for outstanding accomplishments in literature, medicine, physics, chemistry, and peace. His will was bitterly contested but remained unbroken. The first Nobel Prize in Physics was awarded to Röntgen in 1901. That year and the next Marie Curie, Pierre Curie, and Henri Becquerel were nominated by Charles Bouchard, a doctor with lifetime nominating rights, but in 1902 the physics prize went to Hendrik Antoon Lorentz and Pieter Zeeman for their research into "the influence of magnetism upon radiation phenomenon." This was disappointing because Pierre had laid most of the groundwork for these studies. The following year, in a stunning example of what it was to be a woman in science, a vicious sexism ripped away all pretense that Marie Curie might be accepted as an equal.

Four influential scientists collaborated on an official letter nominating Pierre Curie and Henri Becquerel for the 1903 Nobel Prize in Physics. Madame Curie was not mentioned. The letter contained a distorted account of the discovery of polonium and radium. It asserted that these two men, competing against foreign rivals, had "worked together and separately to procure, with great difficulty, some decigrams of this precious material." This in spite of the fact that Marie Curie's amazing discoveries were known throughout the scientific community and that three of the four men who signed the letter had been involved in her work and knew full well to whom the credit belonged. The most shocking of the four was Gabriel Lippmann, whom Marie had deemed a close friend and advisor. Lippmann, however, had regarded Marie as an impoverished young student, not as a potential competitor.

There was speculation that Becquerel had influenced the letter in order to cast more credit on himself. One member of the Nobel science committee, Magnus Gösta Mittag-Leffler, a famous mathematician and chief editor of *Acta Mathematica*, believed that women in science were unappreciated and deplored Madame Curie's omission from the nominating letter. To test the waters, he wrote privately to Pierre apprising him of the situation. Pierre responded that if this nomination was serious he could not accept the prize unless the Nobel committee included Madame Curie. Armed with Pierre's reply, Mittag-Leffler exerted his considerable influence to urge that Marie Curie's name be added to the letter of nomination. Certain adversarial committee members claimed this was impossible since the nomination letter had already been filed. It was then that Charles Bouchard reminded the committee that this was not strictly true since he had included

Marie in his nominations for the Nobel Prize both in 1901 and 1902. By now the politics of the committee had grown so fraught that at last they added Madame Curie's name to the award. By this technical fluke, she was credited with "opening up a new area of physics research" and for her part in "the most magnificent methodical and persistent investigations." With a glancing reference to the work of Rutherford and Soddy, the Nobel Prize report noted that although the Curies "have sometimes been overtaken by other scientists . . . this can by no means diminish the honor due them for the first discovery of the phenomenon of radioactivity."

Today a Nobel Prize is celebrated, but at the time these science prizes were known only within the scientific community and valued for the 70,000 gold francs given to the recipient. The money that would accompany the award seemed to be what most excited Marie, and she wrote Bronya of how it would further their research. However, Becquerel received his 70,000 gold francs but Marie and Pierre, as if they were one person, shared the same amount.

In November 1903, the Curies received formal notice that they had won, and an invitation to receive the prize on December 10 in the presence of King Oskar II. To the surprise of the scientific community, the Curies accepted the prize but declined the trip to Sweden, the first recipients to do so. The problem, well hidden at the time, was Marie's condition. Once again, her pattern of recurring depression had asserted itself. The previous summer, although five months pregnant, she had taken a bicycle trip with Pierre, who had emphasized how much he needed her company, oblivious to the danger such a strenuous trip might entail. After three weeks of constant biking, Marie suffered a miscarriage. When she could obsessively throw herself into work, she could keep going, but now the arduous labor was behind her. She was depleted physically and had not yet had time to mourn the loss of her father or her unborn child. The following week she took to her bed, speaking little, eating less, ignoring Irène, rousing herself only to teach her classes at Sèvres.

Henri Becquerel wearing the green brocade costume and sword of a member of the Academy of Sciences.

Henri Becquerel appeared alone to accept the prize. At an evening reception he wore a green brocade and gold-embroidered waistcoat, medals gleaming across his chest, a sword at his side connoting his membership in the Academy of Sciences. At the Nobel Prize ceremony, Dr. H. R. Törnebladh, president of the Royal Swedish Academy of Sciences, emphasized Becquerel's contributions and three times solely credited him with the discovery of radioactivity:

> The promise for the future stemming from Becquerel's discovery seems near full realization. . . . Professor Becquerel, the brilliant discovery of radioactivity shows us human knowledge in triumph, exploring Nature by undeflected rays of genius that pass through the vastness of space. Your victory serves as a shining refutation of the ancient dictum, ignoramus–ignorabimus, we do not know and we shall never know.

The Curies were relegated to their "comprehensive and systematic research into this topic."

Gender bias once again became glaringly obvious when Dr. Törnebladh concluded his speech by observing, "The great success of Professor and Madame Curie . . . makes us look at God's word in an entirely new light: It is not good that the man should be alone; I will make him an help meet for him." The biblical reference to Eve, the evil temptress of man, underlined the prevailing attitude. Although like her husband she too was "Professor Curie," she would in this speech and even to the present be referred to as "Madame Curie." But against all odds, Marie had become the first woman Nobel laureate and for thirty-two years (with the exception of her second Nobel Prize) would remain the only woman to receive this honor, until it was awarded to her daughter Irène Joliot-Curie in 1935.

The 1903 Nobel Prize in physics made no mention of the discovery of the elements radium and polonium. It is commonly thought that this omission was meant to leave the door open for a future Nobel Prize in chemistry. In the presentation speech a more likely reason emerges. When Törnebladh said, "We have found a new source of energy, for which the full explanation is not yet forthcoming," he meant, What if Marie's elements did not exist?

Perhaps the German scientist Willy Marckwald, a professor of chemistry at the University of Berlin, had sown the seeds of doubt. A few weeks before the prize was given, Marckwald read a paper by Madame Curie in which she observed that polonium, although a great deal more radioactive than radium, was more difficult to isolate because it evaporated in six days unless kept in a sealed container. Marckwald wrote that this was because polonium was not a new element, as she claimed, but a compound. He added that in the course of his investigation, however, he had isolated a new element and named it "radiotellurium." Marie reacted to this criticism by going directly to the laboratory, where she began a study of Marckwald's so-called element. Then she wrote a paper in German, so it could not fail to be noticed in Berlin, in which she asserted that he had misread her meaning and furthermore the element he described was identical to polonium in all respects.

Somehow Markwald's attack galvanized Marie. She threw off the last vestiges of depression and plunged back into work, conducting an exhaustive study of polonium and radiotellurium that took more than two years to complete. Polonium, Marie found, fit into Mendeléev's periodic table at number 84, right after bismuth. She set its weight at approximately 212 and the half-life of its most common isotope at 140 days, thus irrefutably proving Marckwald wrong.

CHAPTER 11

"The Disaster of Our Lives"

Marie and Pierre were about to make a Faustian bargain. In the past, they had fought prejudice, neglect, cynicism. Now a newfound celebrity brought with it a cornucopia full of their greatest desires. In return, all that had been most meaningful in their lives began slipping away. The Curies, who had "dreamed of living in a world quite removed from human beings," were besieged by the press. They sensed that this recognition might bring them the rewards they sought, but they were not prepared for the frenzy that followed. The Nobel Prize, only in its third year, had attracted little attention, especially in the sciences. But here was a human-interest story made for the press. As in a fairy tale, Marie was depicted as a beautiful, poor immigrant, a Cinderella who lived in a garret. Cold and hungry, she studied deep into the night. Then she met her Prince Charming in the person of Pierre Curie. Finally, after years of toil in miserable conditions, she discovered a luminous, magical substance that might prove to be a panacea for the world's ills.

The publicity was concentrated mostly on Marie, but after a childhood of discipline and suppressing her emotions, she received fame with equanimity. Pierre, the most private of men, called it "the disaster of our lives." He was troubled that, although the Nobel Prize was for the discovery of radioactivity, it was radium itself that fascinated the public. He wrote a friend:

> I have wanted to write to you for a long time; excuse me if I have not done so. The cause is the stupid life which I lead at present. You have seen this sudden infatuation for radium, which has resulted for us in all the advantages of a moment of popularity. We have been pursued by journalists and photographers from all countries of the world; they have gone even so far as to report the conversation between my daughter and her nurse, and to describe the black-and-white cat that lives with us. . . . Finally, the collectors of autographs, snobs, society people, and even at times, scientists, have come to see us . . . and every evening there has been a voluminous correspondence to send off. With such a state of things I feel myself invaded by a kind of stupor. And yet all this turmoil will not perhaps have been in vain, if it results in my getting a Chair and a laboratory. . . .

Pierre, thinking that publicity would bring scientific support, granted interviews but clearly this was painful for him. He often answered questions with a nod of his head or shrug of his shoulders and frequently looked at his watch as if to say, "This is a waste of time," Marie wrote. "From early youth Pierre [found it] necessary . . . to concentrate his thoughts with great intensity upon a certain definite object. . . . It was impossible for him to modify the course of his reflections to suit exterior circumstances." Not so for Madame Curie, who stoically and politely received all callers.

Today the walls of the Curie Institute are lined with photographs of Marie and Pierre in various scientific poses. The most famous one is of Madame Curie holding aloft beakers of radium bromide. "You know, she posed for that picture," Hélène Langevin-Joliot remarks. On close inspection the photograph seems forced, Marie's arms are awkwardly extended, her eyes are glazed rather than concentrated. Yet the Curies had judged correctly, fame had brought long-sought-after rewards—the Davy Medal of the Royal Society of London,

The Marie Curie myth begins to build.

twelve honorary doctorates, academy memberships in several countries, highly paid invitations to lecture. After being rejected by the Academy of Sciences, Pierre (but not Marie) was now a member. The president of France, Émile Loubet, visited the Curies and was photographed in their so-called laboratory.

It had become an embarrassment to France to find its most famous scientists occupying inferior positions. Impelled by public opinion, spurred on by the press, the director of the Academy of Sciences petitioned the Chamber of Deputies (the French Parliament) to create a new science chair at the Sorbonne and name Pierre as its occupant. The chair came with a salary of 10,000 francs, but made no provision for a laboratory. A newly influential Pierre, with public support behind him, declined the chair. This was followed by negative publicity so telling in its impact that the Sorbonne recanted and promised Pierre a fully equipped laboratory and three assistants of his choosing. Madame Curie was to be named head of research.

The Curies' dream of an unencumbered life in science at last seemed obtainable, but Pierre's ambivalence and distress were evident when he wrote to Georges Gouy, "As you have seen, fortune favors us at this moment; but these favors do not come without many worries. We have never been less tranquil than at this moment. There are days when we scarcely have time to breathe. . . ."

In the face of their celebrity, the work that had brought such happiness had diminished:

We continue to lead the same life of people who are extremely occupied, without being able to accomplish anything interesting. It is now more than a year since I have been able to engage in any research, and I have no moment to myself. Clearly I have not yet discovered a means to defend ourselves against this frittering away of our time which is nevertheless extremely necessary. Intellectually, it is a question of life or death.

Pierre struggled on. He experimented with the force of gravity on such radioactive materials as radium and thorium and studied the radioactivity of several thermal water sources. His main efforts, however, went toward developing medical uses for radium. The commonly held perception is that radium treatments changed the course of medicine directly following Marie's discovery, but until the 1930s such applications were extremely limited because of the scarcity and high price of the pure radium required in performing radiation treatments. The *British Journal of Radiology* stated that initially Marie Curie produced only one grain (0.065 gram) of radium from two tons of pitchblende residue. With the help provided by the Central Society's facilities, by 1904 it was considered an achievement when the yield was four grains of radium (0.26 gram) per ton of this ore.

At this time the medical use of radium was problematic, but bogus applications of radium substances were growing at an astonishing rate into a multimillion-dollar industry. The vast energy emitted by pure radium allowed it to be diluted up to 600,000 times by such substances as zinc sulfide, zinc bromide, or other bromides and still retain its power. The radium craze was not to abate for over four decades. Products containing radium were perceived as a cure for real and imaginary illnesses and as a novelty for society. George Bernard Shaw wrote, "The world has run raving mad on the subject of radium, which has excited our credulity precisely as the apparitions at Lourdes excited the credulity of Roman Catholics."

Minute dilutions of radium were added to tea, health tonics, face creams, lipsticks, bath salts, costumes that glowed in the dark, and so forth. La Crème Activa, purported to contain radium, was guaranteed to keep skin looking young. Curie Hair Tonic guaranteed no loss of hair. A bag containing radium worn near the scrotum was said to restore virility; a Cosmos Bag was strapped to the waist for arthritis. Radium toothpaste was said to preserve and whiten teeth, a radium inhaler to increase the vigor and enrich the blood. A doctor calling himself "Alfred Curie" marketed Créme Tho-Radia. His advertisements showed a beautiful blonde woman with flawless skin bathed in blue light. According to Hélène Langevin-Joliot, Marie was so offended by this appropriation of the Curie name that she asked a lawyer to write him to desist. Nevertheless he continued.

One could buy a Revigorator—a flask lined with radium to be filled with water each night to drink the following morning. Radithor, a drink containing one part radium salts to

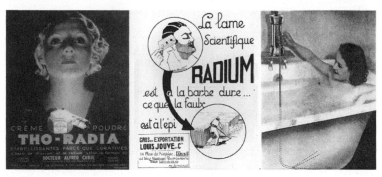

Radium advertisements promising to fulfill your dreams.

60,000 parts zinc sulfide, was said to cure stomach cancer, mental illness, and restore sexual vigor and vitality. An American industrialist, Eben Byers, drank a bottle a day for four years, at the end of which he died in excruciating pain from cancer of the jaw as his facial bones disintegrated. The famous American Follies Bergère dancer Loie Fuller became infatuated with Marie and her discovery and wrote requesting some radium to create a costume. When Marie refused, Loie came to the Curies' house and performed a dance, her body lit by electric lights colored by blue cellophane filters—the nearest she could come to a radium effect. Soon, in Paris, New York, and San Francisco, theater and nightclub reviews featured women invisible but for the glowing radium paint on their costumes.

Radium had become the pet substance of high society. In fashionable drawing rooms, society hostesses paid exorbitant prices to have so-called experts lecture on radium. Upper-crust men and women carried in their pockets or purses glass vials containing tiny particles of radium bromide. When Pierre heard of this, he wrote a paper warning about the danger of burns, but perhaps flirting with danger was alluring to affluent people as was the frequent use of morphine injections and cocaine.

A device called a spinthariscope, invented by Sir William Crookes to detect radioactivity, soon became favored by society. A round brass tube about two inches long housed a tiny mirror and a transparent screen covered with zinc sulfide. A particle of radium bromide, diluted to about one part pure radium in six hundred thousand was placed inside. This was no toy. It demonstrated the fierce energy of radioactivity. On the screen one saw a multitude of scintillations, like minute shooting stars, as the substance decayed. Crookes patented his sphintariscope. Everywhere profits poured in.

The Central Society had made it possible for Madame Curie to have enough processed pitchblende residue, enough staff, and enough money to discover radium, but now the allure of greater rewards beckoned the Curies. In 1904, they were persuaded by Armet de Lisle, a clever businessman with a flourishing quinine factory, to switch their allegiance to him. They signed a new contract which guaranteed the production of radium on a larger scale, and a laboratory in which to work. De Lisle did not rely on scientific data but rather on the new-found status and celebrity of the Nobel Prize–winning Curies. He built a factory to manufacture "radium salts" in Nogent-sur-Marne. It employed eighty-five people.

De Lisle published a journal, Le Radium, ostensibly to inform the public about this miracle substance and to point out its uses in science and industry, thereby fanning the flames of radium's popularity. The price of radium was now escalating beyond any expectation. By 1904, a gram of radium cost 750,000 gold francs (110,710 U.S. dollars today). The slag heap in the Austrian forest had turned into a mountain of gold.

As supplies of radioactive ore began to diminish, de Lisle placed advertisements in Le Radium encouraging his readers to prospect for French radioactive ore. When none turned up, he financed a successful mission to faraway Madagascar and processed the ore at his factory. De Lisle's ambition was limitless. He then published a magazine promoting the use of radium salts for agriculture. Experiments were conducted at the factory on the effect of radiation on hundreds of grains and on the germination of plants, some of which produced strange, vibrant flowers. The factory sold fertilizers guaranteeing to double crop yields. When Marie measured the radioactivity of these fertilizers, she found almost none. To stretch his profits further, de Lisle hired two of the Curies' former assistants, Jacques Danne and Frédéric Haudepin, to develop "efficient radium salts," which meant those containing the least pure radium. Within a short time, the factory routinely sold radium salts that had been diluted 600,000 times.

While the Curies received de Lisle's processed radium for research and royalties on the sales, they were far from rich, for they held no patent on radium or on the process by which it was manufactured. In 1923, when Marie was trying to raise money to continue her research, she wrote a short autobiography in which she stated that they had considered taking these steps and knew they would "sacrifice a fortune" if this was not done, but that she told Pierre that such action "would be contrary to the scientific spirit. . . . If our discovery has a commercial future, that is an accident. Radium is going to be of use in treating disease. . . . It seems to me impossible to take advantage of that." Pierre agreed with her: "No. It would be contrary to the scientific spirit," he said.

The facts, however, reveal a much more complicated attitude on the subject of money. First, could the Curies have patented radium or its process of manufacture even if they had seen how lucrative it was to become, which they did not? In 1899, Pierre had contacted the Central Society simply to facilitate Marie's investigation of radium. At that time, Pierre tried but failed to generate industrial interest in radium, and its medical applications were unknown. The radium fad had not yet been imagined. It was unclear that radium would become valuable.

Hélène Langevin-Joliot feels that it would have been impossible to patent radium itself. As to the treatment process, by 1903 when the Curies began to understand that there was money to be made, marketable radium salts were also being processed in Germany by a different method than that of the Curies. Also, Marie herself varied her methods. There was no standard by which to grant a patent.

In any case, the Curies were conflicted on the subject of money. The public expected scientists to be so dedicated and idealistic that they could live on air. A basic irony was that many of these scientists also subscribed to this view. Unless they came from wealthy families as had Becquerel or had married into wealth, they lived on the edge of poverty, yet many of them were convinced that to make money from their discoveries was crass and immoral. Scientific knowledge was to be freely shared for the good of humanity. Röntgen gave his Nobel Prize money to charity only to die in abject poverty after World War I. J. J. Thomson, one recalls, had convinced Rutherford that it was dishonorable to derive profit from his discoveries.

Since earliest childhood Marie had looked down on those who lacked dedication to their country, to higher ideals, to their work. She belonged to a class of people who prided themselves on intellectual achievements, not on material possessions. This belief armed the Curies against the slights and wounds of those members of the establishment who had shunned and ignored them. Even when there was enough money to live comfortably, Marie, having endured an impoverished youth where the price of a stamp or a scuttle of coal was a matter of concern, remained frugal. Throughout her life she collected string and used discarded cardboard for her scientific calculations. Her worn dresses were refurbished until they became unwearable.

She disdained the world of fashion. When her daughter Eve became a fashionable beauty, Marie's ridicule contained both elements of cruelty and protection. "What sort of new style is this . . . miles and miles of naked back! You run the risk of pleurisy" or, "You'll never make me believe women were made to walk on stilts." Ernest Rutherford summed up this attitude when he said, "We don't have money so we have to think."

"Science is a wonderful thing if one does not have to earn one's living at it," said Albert Einstein. Money, or the lack of it, provided major conflicts throughout Marie's life. When in 1898 the Central Society began selling samples of radium salts to scientists in Germany, Canada, England, Austria, and America, the Curies lent free samples of these precious radium substances to qualified scientists in these countries, as well as in Poland and Iceland. The free Curie sample they gave to Ernest Rutherford, which was three hundred times more active than a German sample he had purchased, enabled him to prove his theory of transmutation. Friedrich Giesel, the director of a German chemical factory, compared a Curie radium sample with that of his own factory and wrote, "It goes without saying that this is why your research is more effective. You can observe phenomena which are not perceptible here."

But as the popularity and price of radium escalated the Curies were not exempt from temptation. Armet de Lisle persuaded Pierre to modify his instruments to sacrifice accuracy for portability, thus making them more salable. Pierre designed carrying cases and patented them, as well as the instruments themselves. His royalties were considerable. In

1904, the Austrian government, which controlled the pitch-blende residue in St. Joachimsthal, financed their own factory to process and sell radium. They allowed the Academy of Sciences of Vienna to buy twenty tons of pitchblende ore. Then they embargoed all other countries but, from 1904 to 1906, allowed the Curies to buy twelve and a half tons of pitchblende residue at a reasonable price; the Curies in turn denied free samples to other scientists. De Lisle was accused of masterminding the Austrian embargo, and the Curies, who felt proprietary about Marie's discovery, were accused of keeping the lion's share of this material and its profits under their control. This, of course, infuriated scientists in other countries who appealed to the Radium Institute of Vienna to lift the embargo, to no avail. Soddy wrote to Rutherford, "I have a shrewd suspicion Curie has lobbied the Austrian government and secured the monopoly of the St. Joachimsthal mine, damn him. . . . The residue is not to be had. They will soon look bad, when they find some in your part of the world." No ore was found in Great Britain, but Americans began mining and found rich sources of radioactive ore. By 1906, the Lockwood extraction plant in Buffalo, which exceeded the production of the de Lisle factory, began exporting radium, thus forcing the Austrian government to lift the embargo. Pierre explained the dilemma of balancing ideals with the necessities of life:

> We must make a living, and this forces us to become a wheel in the machine. The most painful are the concessions we are forced to make to the prejudices of the society in which we live. We must make more or fewer compromises according as we feel ourselves feebler or stronger. If

one does not make enough concessions he is crushed; if he makes too many he is ignoble and despises himself.

I believe that justice is not of this world, and that the strongest system or rather the one best developed from the economic point of view will be that which will stand. A man may exhaust himself by work, and yet live, at best miserably. This is a revolting fact, but it will not, because of that, cease. It will disappear probably because man is a kind of machine, and it is of economic advantage to make every machine work in its normal manner, without forcing it.

CHAPTER 12

"We Were Happy"

Sixteen months had elapsed since the Curies won their Nobel Prize and their life had changed completely. "To tell the truth I can only keep up by avoiding all physical fatigue and my wife is in the same condition. We can no longer dream of the great work days of times gone by," Pierre wrote in the spring of 1905. Though Pierre had observed the death of laboratory animals when exposed to radium, neither of them associated their deteriorating health with this substance. The Nobel Prize required that the recipients present a lecture, which usually took place at the time of the award, but it was not until April that Pierre and Marie had the strength for the forty-hour trip to Stockholm. Pierre alone was asked to speak. He was seated on the dais, she was in the audience. This insult turned out to Marie's advantage since her husband, from the podium, could then give her full credit for her discoveries. In his speech "Radioactive Substances, Especially Radium," he mentioned Madame Curie's accomplishments again and again:

Mme. Curie showed in 1898 that of all the chemical substances prepared or used in the laboratory, only those containing uranium or thorium were capable of emitting a substantial amount of Becquerel rays. We have called such substances *radioactive*. Radioactivity, therefore, presented itself as an atomic property of uranium and thorium.

Pierre pointed out that Marie alone had discovered the radioactivity of these elements and noted that substances such as polonium and radium existed in pitchblende residue "only in the form of traces, but they have an enormous radioactivity," and for the first time, raised the catastrophic possibilities of her discovery. In his closing words, this idealist admitted a balance of good and evil:

> It can even be thought that radium could become very dangerous in criminal hands, and here the question can be raised whether mankind benefits from knowing the secrets of Nature, whether it is ready to profit from it or whether this knowledge will not be harmful for it. The example of the discoveries of Nobel is characteristic, as powerful explosives have enabled man to do wonderful work. They are also a terrible means of destruction in the hands of great criminals who are leading the people towards war. I am one of those who believe with Nobel that mankind will derive more good than harm from the new discoveries.

Ernest Rutherford was less optimistic and noted that, with all the energy generated by radioactive change, "some fool in a laboratory might blow up the universe unaware." And Soddy wrote:

It is probable that all heavy matter possesses—latent and bound up with the structure of the atom—a similar quantity of energy to that possessed by radium. If it could be tapped and controlled what an agent it would be in shaping the world's destiny! The man who put his hand on the lever by which a parsimonious nature regulates so jealously the output of this store of energy would possess a weapon by which he could destroy the earth if he chose.

He added, "Radium and radioactivity have transformed the earth into a stockhouse stuffed with explosives, inconceivably more powerful than any we know of."

On returning from Sweden the Curies, finally following the French custom, left Paris for the summer and rented a modest cottage on the Normandy coast. Here Marie regained her equilibrium and soon was swimming in the rough water and playing with Irène. She invited her sister Helena Szalay and daughter Hania, who at seven was a year younger than Irène, to come for a visit. She felt they, too, were much in need of a respite. Warsaw was in turmoil. For the first time, Japan had defeated a Western nation in the Russo-Japanese War, which began after Russia failed to honor its agreement to withdraw its troops from Manchuria. The Russian fleet had been destroyed, the country's military humiliated. These events were followed by the smoldering discontent of the countless Russian peasants and impoverished workers whose children had no future. On January 9, 1905, more than a hundred and forty thousand peasants and workers, many accompanied by wives and children, marched peacefully to the Winter Palace in St. Petersburg bearing a petition which described their plight. Tzar Nicholas II gave the order to fire. More than a thousand men, women, and children lay dead and another five thousand injured. The result was mass demonstrations in many cities including Warsaw. Barricades lined the streets; preparations for a general strike were put into place. For Lenin this "was the great rehearsal" for the Russian Revolution of 1917.

In the face of her newfound stardom, of the idolatry of radium that had begun to exceed her own, Marie drew her family close. She could be open and relaxed only with the few she trusted and truly loved. With her sister Helena Szalay she reminisced about their childhood and, as Helena wrote, "our youthful dreams [and] all the pain and disappointment." Marie helped the girls collect shells, admiring each as if it were a jewel. That summer Pierre's legs were so painful and his balance so unsteady that he was unable to walk on the tamped-down coarse sand. Marie confided to her sister that he could not sleep because his back pain was becoming stronger and stronger accompanied by acute attacks of weakness. To Helena's chagrin, Marie burst into tears. "Maybe it is some terrible disease that doctors don't recognize," she said, "Maybe Pierre will never be well again."

When the family returned to Paris in the fall, Pierre, though exhausted and ill, returned to work. At the de Lisle laboratory he worked closely with several doctors to further develop the nascent medical applications for radium. For the first time Marie's single-minded devotion to science had softened. She was once again pregnant and this time vowed to take care of herself. On December 6, 1905, a second daughter, Eve Denise, was born and the expenses for telegrams and a celebratory bottle of champagne were entered in Marie's journal. As the last traces of depression vanished, Marie finally purchased a few new dresses and the Curies saw Eleonora Duse in Maxim Gorky's *The Lower Depths*. They visited Auguste Rodin in his studio and went to the top of the Eiffel Tower. One night Marie dressed in a black grenadine and white chantilly lace gown, wound her hair into a chignon, and hung a dainty gold filigree necklace around her neck. Pierre looked at her and was struck by her beauty. "Evening dress becomes you," he said with admiration, then sighed and added, "But there it is, we haven't got time."

Behind this less strenuous schedule was an unspoken reason: Pierre's general nervousness and health were worsening. His new chair in physics at the Sorbonne had enabled him to resign from the School of Physics, and he chose their close friend Paul Langevin to serve in his place. To fight his "constant enervation and fatigue" he engaged a *préparateur*, but still there were days when he felt too weak to dress himself and nights when his bone pain prevented his sleeping. Marie worried constantly about her husband's health. Several times she said aloud that without Pierre she probably would stop work and he reprimanded her gently saying it was wrong to speak that way. "It is necessary to continue no matter what."

By 1906, the Curies' fame had spread worldwide. For Pierre this was a constant annoyance but Marie seemed refreshed by the recognition and positions that were offered, as money became less of an issue. She relished the time spent with her family. When a stranger admired baby Eve's beauty, she told her with mock seriousness that she didn't know where such beauty had come from since Eve was just a poor orphan. After that, Marie, with a sense of humor that had lain dormant since the carefree year she was sixteen, laughingly referred to Eve as "my poor orphan baby."

In Irène at eight Marie saw the same thoughtful, calm disposition that so pleased her in Pierre. In Marie's journal the observations about her daughters proliferate: skinned knees, scarlet fever, whooping cough, Irène's jealousy of Eve. Her domestic life was expanding. She cooked, sewed, rearranged furniture. Marie wanted Pierre to spend more time with the family and to pay more attention to her, but he expressed his annoyance when Marie preferred staying with the children rather than accompanying him to the laboratory as she had in the past. The worse his mysterious illness grew, the more he felt compelled to drag himself to the laboratory to accomplish all he could. He began new experiments on the effects of radioactivity on thermal waters.

In April of 1906, when Irène was on Easter vacation, the Curies journeyed to St. Rémy-les-Chevreuse. This Easter holiday is often presented as an idyll, a series of glorious snapshots: Irène frolicking in a meadow bursting with spring flowers and chasing butterflies; an excursion to a nearby farm to buy milk; fourteen-month-old Eve on a blanket in the sunshine; Marie leaning her head on her beloved Pierre's shoulder, a sunny bouquet of marsh marigolds he had gathered beside them. They were as one. It was then that Marie thought, in spite of Pierre's ill health, that she had everything a woman could desire and in the future "nothing was going to trouble us. . . . We were happy."

But in the real world this Easter vacation also evoked the problems of everyday life which couples must address. Marie was relaxed, enjoying her family and leisure time, but Pierre insisted on working as hard as he could. The family left for St. Rémy without him. Later she wrote that when "you left for the laboratory . . . I reproached you for not saying goodbye to

me." Pierre joined them a week later. At the end of the vacation, Marie begged him to stay a few days more, but he
refused, saying he needed to return to Paris to work. Marie
was "very unhappy about this," so Pierre stayed for the week-
end and Monday took the late evening train to Paris, carrying
a fresh marsh marigold bouquet.

On Wednesday, April 18, Marie returned with the children
to attend a scientific dinner with Pierre. Thursday morning
the unvarnished realities of day-to-day life once again became
evident. The maid wanted a raise. In response, Pierre told her
that her housekeeping was sloppy. Marie was harried, trying
to organize the girls on their first day back. When she said
that she might take Irène for an outing, he forbade her to do
so as he wanted her to accompany him to the laboratory. He
walked downstairs and called up to her asking if she was going
to meet him at the laboratory. Marie snapped back, "I don't
know. . . . Don't torment me."

Pierre took his umbrella from the front hall stand and
stepped out into a drenching Paris rain. He went directly to
the laboratory, then left at ten to attend a lunch meeting of
the Association of Professors of the Science Faculties, a group
of nonestablishment scientists who nevertheless were becoming influential. Pierre felt a kinship with these men of similar
background and the meeting went well. As lunch concluded,
an expansive Pierre invited the seven-man group to his home
for dinner that evening. Whatever warmth was in that room
was quickly dispelled as Pierre opened his black umbrella and
headed for the offices of *Comptes rendus* to proofread his new
paper. The rain-soaked corner where the Pont Neuf and the
rue Dauphine converged was clogged with a chaotic assem-
blage of traffic—delivery wagons, carriages, cabs, buses, lone

riders, people on foot—all jumbled up at the busiest intersection in Paris.

It happened in an instant. Pierre, who limped noticeably from the then unknown or unacknowledged radiation exposure that caused his leg bones to deteriorate, stepped into the traffic at the very moment a heavily loaded wagon pulled by two Percherons came galloping off the Pont Neuf and into the congested intersection. One horse rushed past Pierre, grazing him on the shoulder. He reached up to hold on to the horse's chest in an attempt to steady his weakened legs. Then both horses reared and Pierre slipped and fell between them. The wagon rolled over him. The front wheels missed him but the left rear wheel crushed his skull. He was forty-nine.

The police identified the body from the cards in his wallet and rushed to the Sorbonne to inform the dean of his department, Paul Appell. Appell and Jean Perrin immediately went to the Boulevard Kellermann house. Doctor Eugène Curie opened the door, looked at their faces and exclaimed, "My son is dead. . . . What was he dreaming of this time?"

Later there would be inquiries and finally Pierre Curie's death would be put down to the dismal weather, the obscuring umbrella, and his inattentiveness. Other contributory causes were not addressed: there was no mention of that pernicious shimmering angel of mercy and of death—radium.

CHAPTER 13

Metamorphosis

At dusk Marie and Irène returned from an outing to Fontenay-aux-Roses. It fell to Paul Appell to tell her of Pierre's death. At first she said nothing. After a long silence, she spoke in a barely audible voice, "Pierre is dead. Dead. Absolutely dead?" After making a few hasty arrangements, Marie walked dazed into the rain-soaked garden and sat on a bench, "elbows on her knees and her head in her hands, her gaze empty. Deaf, inert, mute," as she waited for her husband's body to arrive.

The next morning Marie saw the marsh marigolds Pierre had carried still fresh in a vase on the kitchen table. She looked away unable to bear the sight. Years later, Eve Curie, who was scarcely over a year old when her father died, was to write that Pierre's death marked the defining moment in her mother's life:

> It is commonplace to say that a sudden catastrophe may transform a human being forever. Nevertheless, the deci-

sive influence of these minutes upon the character of my mother, upon her destiny and that of her children, cannot be passed over in silence. Marie Curie did not change from a happy young wife to an inconsolable widow. The metamorphosis was less simple and more serious. The interior tumult that lacerated Marie, the nameless horror of her wandering ideas, were too virulent to be expressed in complaints or in confidences. From the moment when those three words, "Pierre is dead," reached her consciousness, a cape of solitude and secrecy fell upon her shoulders forever. Madame Curie, on that day in April, became not only a widow, but at the same time a pitiful and incurably lonely woman.

With Pierre's death Marie irrevocably closed to the world. Never again would there be a sign of joy.

A few days after Pierre's death, Marie began a diary that she kept for almost a year, in which she recorded her deeply felt emotions so different from the impassive face she presented to the world. The very few scholars who are permitted to read this diary come away knowing the unvarnished Marie Curie, not as the icon she has become but as a complicated, passionate, tenacious, directed, melancholy woman. Although this diary, a beige canvas notebook approximately nine by seven inches, is seventy-three pages long, she used only twenty-eight single-sided pages. One page has been ripped out and page 22 has been cut in half (both at points where she becomes critical of life with Pierre).

It is striking that the majority of the diary entries address Pierre as if he were present—striking and odd—until one realizes that the Curies, especially Pierre, believed in spiritualism, a basic tenet of which is the ability to communicate with those who have "passed over." William Crookes, the chemist and inventor of the Crookes tube and the spinthariscope, described the experiments he and some of the most respected scientists of the day conducted with mediums and other spiritualists, in which they applied "crucial tests [using] carefully arranged apparatus, in the presence of irreproachable witnesses." He documents séances with various mediums who he believes can communicate with the dead. Crookes asserts that, guided by "intellect . . . cold and passionless," he has discovered "a new truth." In Crookes's book, *Researches in the Phenomena of Spiritualism*, he states that, after conducting exhaustive scientific investigation, the "Spiritualist phenomena that cannot be explained by any physical law at present known, is a fact of which I am as certain as I am of the most elementary fact of chemistry."

This was a time when many scientists had begun to explore an invisible world and they were convinced that there was a scientific explanation to be found to confirm this belief. Henri de Parville wrote of the Curies' Nobel Prize in *Le Correspondant*, "Everything around us is radiations . . . luminous radiation, caloric, electric, sonorous. . . . Why doubt telepathy, the influence from a distance of thought on thought? The rays which escape from the nervous cells are most capable of exciting other nervous cells from afar." At the dawn of the age of atomic physics, this phenomenon was no stranger than the discoveries being made every day such as electricity, radio waves, magnetism, Röntgen's X-rays, Becquerel rays, and the fierce radioactivity produced by Marie Curie's radium and polonium. In a world where messages were being transmitted invisibly by means of the telegraph, spiritualists came to believe that if this was possible why not a Spiritual Telegraph with which one might communicate with the dead? Crookes declared that this was "authentic communication" through a "psychic force."

The Curies along with their circle of scientist friends—including Crookes; Jean Perrin and his wife, Henriette; Georges Gouy, and Paul Langevin—explored spiritualism as did Pierre's brother Jacques, who was a fervent believer. Pierre and Marie attended many séances, most notably with the Italian medium Eusapia Paladino. They regarded these séances as "scientific experiments" and took detailed notes. The historian Anna Hurwic wrote that the Curies "thought it possible to discover in spiritualism the source of an unknown energy that would reveal the secret of radioactivity." Pierre felt Paladino worked "under controlled conditions." After a séance at the Society for Psychical Research—where in a brightly lit room "with no possible accomplices" he watched as tables mysteriously lifted into the air, objects flew across the room, and invisible hands pinched and caressed him—he wrote Georges Gouy, "I hope we are able to convince you of the reality of the phenomena or at least some of them."

A few days before his death Pierre had written of his last Paladino séance, "There is here in my opinion, a whole domain of entirely new facts and physical states in space of which we have no conception." In 1910, four years after Pierre's death, when Marie was rejected by the Academy of Sciences, Henri Poincaré wrote that Pierre's spirit had come to Marie and tried to comfort her by saying, "You will be elected next time."

In her extreme psychic pain, Marie in her diary seems to speak to her late husband as a spiritualist would. She

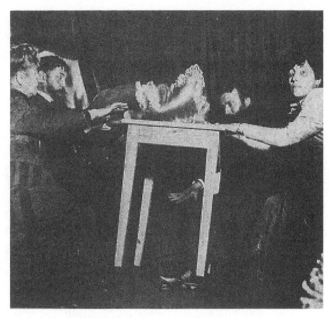

Eusapia Paladino conducts a spiritualist séance attended by scientists, in 1898.

addresses Pierre directly and assures him that she kept his funeral intimate and simple and avoided the "noise and ceremonies you hated." Then she explains,

I put my head against [the coffin.] . . . I spoke to you. I told you that I loved you and that I had always loved you with all my heart. . . . It seemed to me that from this cold contact of my forehead with the casket something came to me, something like a calm and an intuition that I would yet find the courage to live. Was this an illusion or was this an accumulation of energy coming from you and condensing in the closed casket which came to me . . . as an act of charity on your part?

Two days later, in what might be a spiritualist communication, she tells Pierre directly of the idyllic times as they lay in bed "pressed up against one another as usual" and adds, "I sometimes have the absurd idea that you are going to come back. Didn't I have it yesterday, when hearing the sound of the front door closing, the absurd idea that it was you?"

Marie, who sought perfection in every aspect of her life, reacted to her husband's death with self-flagellation and guilt. In her diary, she chastises herself for not going to the laboratory with Pierre that final day when he had wanted her there. Instead she had taken the day off with Irène against Pierre's "expressed wishes." She regrets reproaching him that he was not attentive enough to her and their family. Her last sharp and careless words to her husband bring her immense pain: "When you left, the last sentence that I spoke to you was not a sentence of love and tenderness. . . . Nothing has troubled my tranquility more." She even feels guilty for laughing at a funny word that Irène uses.

On the Sunday after Pierre's funeral, instead of remaining in the comforting arms of her family and friends, Marie retreated to the laboratory. In her diary she tells the dead Pierre of her deep desolation: "Sunday morning after your death, I went to the laboratory with Jacques. . . . I want to talk to you in the silence of this laboratory, where I did not think I could live without you." She continues, "I tried to make a measurement for a graph on which each of us had made some points, but . . . I felt the impossibility of going on. . . . The laboratory had an infinite sadness and seemed a desert. It seems at one moment that I feel nothing and that I can work and then the anguish returns."

That same Sunday morning she began a workbook in

which she documented her experiments. These workbook notations (written simultaneously with her passionate, desperate diary entries) are unemotional, detailed, analytical, and coldly intellectual. In the workbook her comments on her inability to work on a graph without Pierre take on a different interpretation as she notes her experiments and her difficulty in trying to arrange the results on a graph that they had devised together that included specific measurements, dates, and length of time.

Over the next ten months in her scientific workbook she describes the equipment she uses in various experiments, often accompanied by sketches drawn with an artist's talent. Frequently, she repeats the same experiment for several days with minute adjustments. To underscore Pierre's accomplishments for posterity, Marie devoted much of her time to amplifying his experiments on radioactivity and the invisible rays it transmits through air to nearby substances. She verified his findings on the force of gravity on such radioactive materials as radium and thorium and completed Pierre's half-finished book on these topics. Marie, however, refused any credit for this six-hundred-page book. She then edited *Works of Pierre Curie,* a compendium for which she wrote the introduction. Even fifteen years later, she was still paying homage to Pierre when she wrote *Radiology and the War,* listing herself as "Madame Pierre Curie." Marie wrote in her diary, "I live only for your memory and to make you proud of me."

The contrast between Marie's workbook and her diary provides vivid examples of her compartmentalized personality. In her diary the first sign of spring and a new awakening are seen as an open wound so raw that she contemplates death as a welcome release. She tells Pierre, "I walk as if hypnotized without attending to anything. I shall not kill myself, I have not even the desire for suicide. But among all those vehicles isn't there one to make me share the fate of my beloved?" She confides to Pierre, "I want to tell you that I do not like the sun and the flowers anymore, looking at them makes me suffer. I feel better in dark weather, like on the day of your death, and if I do not feel hatred toward the fine weather it is because my children need it. . . . I spend all my days at the laboratory, that is all I can do. I am better there than anywhere else." And shortly thereafter, "the house, the children and the laboratory are my constant concerns." The extent of the damage is evident.

On May 11, 1906, less than a month after Pierre's death, Georges Gouy and Paul Appell arranged for his widow to be offered a national pension. Marie refused it. Gouy then suggested to those in power that she assume Pierre's duties at the Sorbonne. In her diary she wrote:

They have offered that I should take your place, my Pierre. . . . I accepted. I don't know if it is good or bad. You often said to me that you would have liked for me to teach a course at the Sorbonne. Also, I would like at least to make an effort to continue your work. Sometimes it seems to me that that's the way it will be easiest for me to live, other times it seems to me that I am crazy to undertake it.

It would take two more years for the Sorbonne to officially recognize her position. By then Madame Curie was so famous that it was inevitable that she be named a full professor in charge of the chair created for her late husband. In so doing, Marie became the first woman to obtain this position in the history of the Sorbonne.

The famous widow's initial lecture in Pierre's place was scheduled at 1:30 on the afternoon of November 5, 1906. By 10 A.M. hundreds of people had lined up in front of the doors of the Sorbonne's physics lecture hall. Marie's students from Sèvres were to be preseated as were many scientists. Dr. Eugène Curie brought Irène. The child clasped his hand tightly. The hall's capacity officially was one hundred and twenty, but when they opened the doors at 1:15 P.M., several hundred people rushed into the room: journalists, photographers, ladies and gentlemen of society, students, gawkers. It seemed all Paris was there. Most awaited Marie's tears and tributes to her late husband. The crowd stared at the two wide-entry doors at the front of the room, directly behind the rectangular bench where Pierre had lectured and demonstrated his experiments. Virtually unnoticed, Marie slipped into the lecture hall through a back door and only when her black-clad figure reached the table did the audience, suddenly aware of her presence, burst into applause. When the ovation subsided, Marie began to speak in an icy constricted voice. There was no trace of emotion.

When we examine our recent progress in the domain of physics, a period of time that comprises only a dozen years, we are certainly struck by an evolution that has nourished fundamental notions regarding the nature of electricity and of matter. This evolution happened in part because of detailed research on the electrical conductibility of gases, and also because of the discovery and study of the phenomena of radioactivity.

Few noted that her lecture had started at the exact place Pierre Curie's final lecture had left off.

As cool and dry as was the lecture, once again her diary reveals her true feelings. It seems that she is losing her spiritualist belief that Pierre is nearby and aware of her plight. On returning home she writes:

> Yesterday I gave the first class replacing my Pierre. What grief and what despair! You would have been happy to see me as a professor at the Sorbonne, and I myself would have so willingly done it for you, but to do it in your place, my Pierre, could one dream of a thing more cruel. And how I suffered with it, and how depressed I am. I feel very much that all my will to live is dead in me, and I have nothing left but the duty to raise my children and also the will to continue the work I have agreed to. Maybe also the desire to prove to the world, and above all to myself, that that which you loved so much has some real value.
>
> I also have a vague hope, very weak alas, that you perhaps know about my sad life and the effort and that you would be grateful and also that I will find you perhaps more easily in the other world if there is one. . . . That is now the only preoccupation of my life. I can no longer think of living for myself, I don't have the desire nor the faculty, I don't feel at all alive any more nor young, I no longer know what joy is or even pleasure. Tomorrow I will be 39. . . .

CHAPTER 14

"My Children . . .
Cannot Awaken Life in Me"

Many women have been inspired by Madame Curie, this brave woman who defied the strictures against her sex. She has been lauded as proof that women can do it all—and perfectly. The perception abides that, in addition to having a spectacular career, she was a model mother for her two daughters, far ahead of her time in emphasizing the importance of a strong body, a good education, and an unfettered view of life. This too has become part of the Curie legend. The facts, as usual, are more complicated. Marie Curie certainly performed her "duty." Was that enough?

In 1937, three years after Marie's death, her younger daughter, Eve, wrote an award-winning biography of her mother. "I'd never written a book before. I wasn't sure I could do it, but I needed to write this book because it was inevitable that people would write about her, and so few knew her at all." The next thing Eve says is startling to the listener and unwittingly seems to crystallize a great deal of what one wants to know about Marie Curie's relationship with this daughter. "I called the book *Madame Curie* by Eve Curie. I didn't think it right to call it *Marie Curie* by Eve Curie; that would have been too intimate." As if to emphasize the distance between Eve and her mother in her book, Eve often refers to herself in the third person. She writes of her grandfather, Dr. Eugène Curie, "Eve was still too young for a true intimacy to be created between them, but he was the incomparable friend of the elder girl [Irène], of that slow, untamed child so profoundly like the son he had lost."

Eve Curie was only fourteen months old when her father died. She was never to know him or for that matter the mother who had existed before his death. She knew only a morose woman driven by obligation. Shortly after Pierre's death, Marie wrote, "I endure life, but . . . never again will I be able to enjoy it. . . . I will never be able to laugh genuinely until the end of my days." To a former pupil and close friend who felt Marie had been neglecting her she wrote:

> I no longer am able to devote any time to social life. All our friends in common will tell you that I never see them any-more except for business, for questions concerning work or education of the children. No one visits me, and I don't see anyone and I haven't been able to avoid offending some people in my circle and my laboratory who don't find me sufficiently friendly. . . . I have completely lost the habit of conversations without a set goal.

Albert Einstein professed deep friendship for Marie, but described her as "cold as a herring." This was the mother that Eve and Irène grew up knowing. Were it not for Pierre's father, the girls would have endured a bleak life. Dr. Curie, a cheerful, erudite man, had kept the household running while Marie and Pierre became increasingly involved in their quest to isolate radium, determine its weight, and study its radioactivity. He was physically demonstrative and delighted in playing games, joking, and reading to his granddaughters. He took them on outings and explained nature to them as he had to his sons. After her husband's death Marie never allowed Pierre's name to be spoken again. It fell to Dr. Curie to tell the children about their father while their mother was away. He humanized this ghost, relating tales of Pierre's childhood, and how he had a temper like Irène's. He described Pierre and asked, "Can you imagine your father . . . in shorts?"

Marie failed to understand the impact of Pierre's death on nine-year-old Irène. Jean Perrin and his wife, Henriette, lived next door and Irène played with the Perrin children, Aline and Francis. On the day Pierre died Irène was dispatched to the Perrins' and told only that her father had hurt his head. She did not go to the funeral. The following day Marie went to the Perrins' and told Irène of her father's death. The child seemed not to understand but when her mother left she burst into tears. Marie noted that Irène never spoke of her father's death and wrote, "She will soon forget completely." Marie was so cut off from feeling that she was completely unaware of her daughter's anguish.

For a long time after that Irène became angry and anxious if her mother left even for a short time. She woke up from a nightmare and asked pitifully, "Mé [her word for her mother] isn't dead too?" Marie noted her daughter's behavior but attributed no special significance to it; it did not occur to her that Irène suffered in silence. She wrote of her, "She doesn't speak of her father. . . . She no longer seems to be thinking about it, but asked for the picture of her father that we had taken from the window of her bedroom."

Marie's pattern of insensitivity repeated itself when, five years later, Dr. Curie died. Eve, then almost seven, maintains that Irène suffered the most. "I was very young, but Irène was fourteen. My grandfather raised her from the time she was a baby and they had a close bond, a great bond. Irène was desolate when first father died and then her beloved grandfather, who had been everything to her." Once again, Marie took little notice of the pall cast over her household.

The laboratory had become Marie's safe harbor, the one place where she could endure life without Pierre. She drove herself relentlessly, often working until two or three in the morning, and returning to the laboratory at eight that same morning. One of Eve's early memories is of her mother fainting and crashing to the floor. When Marie reached the saturation point of nervous exhaustion, she would retreat to her bed and allow no one to see her. Eve wrote of these episodes that Pierre's brother "Jacques and Marie's brother and sister, Jozef and Bronya, observed with terror the movements of this black-robed woman, the automaton Marie had become: Stiff, absent-minded, the wife who had not joined the dead seemed already to have abandoned the living."

Marie wrote, "I have tried to create a great silence around me." Eve recalled that her mother "would not allow anybody to raise their voice, whether in anger or in joy." Her own voice became so muted, "it could hardly be heard." The cruelest punishment she inflicted on Irène was simply not to speak to her for days at a time. The girls tried desperately to reach and please their austere mother. Eve's and Irène's childhood letters are full of love and longing. When they were sent away for extended trips, Irène writes about how healthy she is, how she is learning her lessons, how much she enjoys mathematics. Marie, who in childhood had never received a mother's caress, rarely gave one, yet this frozen soul kept hidden every letter from her daughters, starting with the most childish scrawl, bound up with confectioner's ribbons. The letters were found after her death, a tacit acknowledgement of a love that could not overtly respond to love.

At a time when upper- and middle-class norms emphasized that women were "the weaker sex," when exercise, higher education, and participation in the world of business and politics were discouraged in favor of domestic skills, Marie created her own rules. She decided country living would be more healthful and moved her tattered household to a house in Sceaux, where Pierre was buried, although it meant an extra half-hour commute on a crowded train. In all seasons she saw that her daughters exercised daily, both at a gymnasium and at home. There were bicycle trips and instruction in swimming. She sent them to Poland, where they hiked and learned to ride horseback. Marie hired Polish governesses to teach the girls her native language.

Regular school seemed "barbarous" in their curriculum for women and their arbitrary restrictions so, with time and effort, this concerned mother organized a group of professors from the Sorbonne, who were her friends and who also had young children, to teach Irène and Eve and seven of their own children. They learned chemistry from Jean Perrin, mathematics from Paul Langevin, as well as "literature, history, living languages, natural science, modeling and drawing." On Thursday afternoons in an unused classroom at the Sorbonne, Madame Curie "taught the most elementary course in physics that these walls had ever heard."

This education lasted only two years before the overwhelmed parents abandoned it, but it demonstrated how much young minds could absorb. Perhaps it was a mixed blessing: In her eighties, Aline Perrin recalled those two years: "It was very good for Irène and my brother, Francis, as they were gifted. But for me, it was too much. Those great scholars working with a little girl. Oh no, really! It didn't make sense."

The admiration women afford Madame Curie for her views overlook the most essential nourishment of childhood: She wrote of her daughters, "They are both good, sweet and rather pretty. I am making great efforts to give them a solid and healthy development. . . . I want to bring up my children as well as possible, but even they cannot awaken life in me." All she had left to give was effort and obligation—nothing more.

The children spent summer vacations by the sea in the care of relatives and governesses. Although Marie carefully monitored their development the time spent with them was scant. From the time of Pierre's death, Irène seemed designated by her mother to fill the gap. Marie had written in her diary to the dead Pierre, "I said it to you often, that this daughter, who promised to resemble you in her grave reflection and calm, would become as soon as possible your companion in work."

As Marie's father had done, she mailed math problems to Irène during their long separations. At eleven, Irène was doing advanced mathematics and wrote apologetically, "I've forgotten a little what you have to do to get the derivation of a radical and of the two numbers that divide it." Eve Curie maintains that her mother was equally interested in both daughters, but in Marie's personal journal one can see that

Irène is the favored child, one that is marked to share her mother's life in science. Her entries often write of Irène's excellence and of Eve's "doing well."

The mother to whom Irène clung dictated that all fear must be conquered. Marie, who took every slight to heart but showed the world an impassive face, did not credit her daughter with the same trait. Irène was a child without a childhood, thrust into the role of her father as over the years she became her mother's confidant and co-worker. By thirteen Irène was traveling alone, spending long periods of time living with Marie's close friends Émile Borel, the famed mathematician, and his beautiful young wife, Marguerite, as well as with the Perrins, while her mother worked, traveled, lectured, and retreated into depression seeing no one. During those periods Irène often wrote her mother plaintive letters:

> When it rains, I think that these dark moments spent waiting for light would be much nicer if you were in a chair next to me. And when I see the sun shine in the sky and make beautiful reflections on the water in the streams, I think that everything would be nicer if a sweet Mé were there, near me, to look at them.

At Sevigné junior high school, Irène so excelled in math and physics that she was allowed to teach these subjects to her peers. At fourteen, she passed the first phase of her baccalaureate and finished her first examinations a year and a half later with honors.

Eve moved in another direction: at three and a half this child began to demonstrate what her mother termed "astonishing musical abilities." Marie had little affinity for music or for that matter any of the arts. But when Eve was twelve, through her connections, Marie secured an evaluation by the great Polish pianist, Ignacy Jan Paderewski, who confirmed that Eve had "exceptional ability." This sparked an unexpected burst of emotion and perhaps relief from her mother. Could a Curie be less than exceptional? Marie, ever careful with money, splurged on a grand piano—a massive instrument of mahogany with ivory keys and swirled legs that is still in use in the house in Sceaux, which was inherited by Marie's granddaughter Hélène.

For open, joyous, emotional Eve, it was hard to be excluded from access to her mother. She recalls Irène and their mother discussing formulas and experiments. Eve tried to make the best of being left out. She imagined that certain algebraic terms employed by her mother and sister, "Bb' and Bb^2 . . . were really charming babies who Marie and Irène Curie were forever talking about. [In French, Bb is pronounced bébé, meaning "baby." The symbols ' (prime) and ^{2} (squared) have the same meaning in French as in English.] . . . But why prime babies? and square babies?" By transmuting this technical conversation to fantasy she comforted herself.

Like Marie, Irène was indifferent to clothes—cheaper and less were better for both. Eve loved clothes and even as a child tried to brighten the bleak rooms with bits of fabric and colorful drawings. Eve writes of her interaction with her mother as if she were a detached observer: "If Eve was going out for dinner, Madame Curie would come into her room, lie down on the divan and watch her dress." Commenting on Eve's makeup Marie would say, "I think it's dreadful. . . . You torture your brows, you daub at your lips without the slightest

useful purpose. . . . I like you when you're not so tricked up."
In a mixture of regret and admiration Eve wrote,

> The struggle against sorrow, active in Irène, had little suc-
> cess in my case: in spite of the help my mother tried to give
> me, my young years were not happy ones. In one single sec-
> tor Marie's victory was complete: Her daughters owe to her
> their good health and physical prowess, their love of sports.
> Such is, in this matter, the most complete success achieved
> by that supremely intelligent and generous woman.
>
> It is not without apprehension that I have striven to grasp
> the principles that inspired Marie Curie. . . . I fear that they
> suggest only a dry and methodical being, stiffened by prej-
> udice. The reality is different. The creature who wanted us
> to be invulnerable was herself too tender, too delicate, too
> much gifted for suffering. She, who had voluntarily accus-
> tomed us to being undemonstrative, would no doubt have
> wished, without confessing it, to have us embrace and
> cajole her more. She, who wanted us to be insensitive,
> shriveled with grief at the least sign of indifference.

Even in this exculpation, as in much of her biography, Eve
refers to her mother as if they were not related. The distance
is ever present.

CHAPTER 15

"The Chemistry of the Invisible"

" **M**adame Curie since the death of her illustrious husband has not accomplished anything by herself. . . . She has stood by the wayside while others are unraveling the mysteries of the atom," wrote a well-known academician in 1910 in the newspaper *Le Temps.* "The recent work of the Curie Laboratory illustrates . . . the absence of true innovation . . . it is busy work," asserted the tabloid *Excelsior.* But those who compare Madame Curie's work after her husband's death with that of the well-known atomic scientists of her time, and find her wanting, miss the point. One should not think of her in terms of the progressive discoveries of the structure and power hidden inside the atom. Madame Curie and her laboratory were dedicated to "medical, biological and industrial research for the peaceful benefit of humanity."

At that time, radioactivity dealt with the spontaneous emission of radiation from such elements as radium, polonium, and thorium as well as the study of the physical and chemical properties of these substances. Marie wrote, "I truly want radioactivity, a science born in France, to develop here." This in no small part was because she could then control its uses. Marie Curie believed that her research could help create a better world. She hated war and found it "senseless" and no doubt remembered her late husband's warning that if radioactive substances fell into the "wrong hands" they could lead to massive destruction.

Once she had assured Pierre's star in the scientific firmament, the laboratory on the rue Cuvier became her own and reflected her goals. She pursued the use of radium for medical treatments and for industrial uses, aided by her connection to the de Lisle factory that processed her materials and manufactured products for these purposes. The factory and her laboratory functioned as a training ground for medical technicians and "in the development of industrial uses for radium-based products." An important but little-noted contribution to science was her tenacious and impeccable work in metrology (the science of weights and measurements). Using an approach that combined both physics and chemistry, she focused on the measurements of radioactive substances. In this she was the best in the world.

In dealing with radioactivity, the question was how could one identify a radioactive element and quantify its energy if one were unable to isolate it? Madame Curie called this challenge "the chemistry of the invisible." In this work her most important ally was André Debierne, the scientist who was so retiring that his forty-year devotion to Marie and his own discovery of the element actinium are rarely mentioned. Debierne negotiated with radium factories, set up facilities, and conferred with other scientists to insure Marie's primacy in the study of radium. He even cared for her and her children when she was unable to do so.

With Debierne at her side, in 1906 she began to concentrate not only on radium but also polonium. Once again she needed processing help and sent Debierne to the factory to request these services. De Lisle complied. Polonium at a near-final stage was prepared for her. She performed the fractionations on-site until this element was finally verified by spectroscopy. The concentration of polonium in the pitchblende residue was 4,000 times less than in radium but its power was even stronger. Research into this substance continued. Debierne's actinium was even harder to identify by spectroscopy, but finally a spectral line established it as a radioactive element. However, because of its scarcity and extremely brief half-life it did not lend itself to further exploration. Marie's superior knowledge of metrology was now essential to atomic research. In 1903, for example, when Rutherford proposed his theory of transmutation, many scientists disputed it and Rutherford admitted that it would be very difficult to prove. Curie's work showed that Rutherford's theory was correct. He wrote:

Apart from the interest of obtaining a weighable quantity of polonium in pure state, the real importance of the present investigations of Mme. Curie lies in the probable solution to the question of the nature of the substance into which the polonium is transformed. . . . It was a matter of very great interest and importance to settle definitively whether polonium changes into lead. . . . The experiment of Mme. Curie and Debierne has settled this question conclusively.

The Curie laboratory became the preeminent institution dealing with the production and certification of radium for industry, medicine, individuals, countries, and administrations. Her employees grew from eight in 1906 to twenty-two in 1910 as well as twenty women scientists who volunteered to work without pay.

In 1907, a Viennese scientist of twenty-nine, a woman who idolized Madame Curie, applied as a volunteer at her laboratory. She was rejected. Her name was Lise Meitner. One of eight children of a well-to-do Austrian lawyer, the family was Jewish but considered themselves fully assimilated. Most had converted to Catholicism or became Protestant as had Lise. Meitner, who was to become both Marie and Irène's scientific rival, was twenty-five when the 1903 Nobel Prize was awarded to the Curies, which inspired her to study radioactivity. She too struggled to achieve a scientific education. Her father encouraged his children to learn science but insisted that Lise be certified to teach French in a female finishing school, which he deemed a proper profession for a woman. This led to what she later termed "my lost years."

Women's education in France, Austria, and Germany usually ended at fourteen. Until then they learned domestic skills to prepare them for marriage and childbearing. Preparatory schools were for boys only. When finally Meitner's father agreed to hire a tutor for her, she was able to complete eight years of schoolwork in two. Women were barred from entering Viennese universities until 1899, but in 1901 she entered Vienna University and in 1905 became the second woman to earn a doctorate in physics. Meitner's initial postdoctoral experiment was with pure radium salts donated to the university by the Curies in gratitude for Austria's allowing them to buy embargoed pitchblende. Meitner demonstrated that the alpha particles that streamed out of this material could be deflected slightly while traveling through matter. It was clear early on that she had a gift for interpreting the work of other scientists and using it as a base for further experiments.

After Madame Curie rejected Meitner (who later said that since Irène was the "princess" of the Institute, her mother wanted no other "good minds"), she accepted an offer from Max Planck, the famous theoretical physicist whose research led to the quantum theory of the atom, to become a volunteer at the Chemistry Institute of Berlin University. Women were relegated to the basement and not permitted upstairs where the laboratories were situated or in the lecture rooms. They were even banned from the bathrooms. Anxious to hear the chemistry and physics lectures, Lise would sneak upstairs and hide beneath the seats. Ernest Rutherford, who had been impressed with Meitner's work, knew that Otto Hahn, his former associate, was looking for a collaborator. Here was a woman who wanted laboratory work, who cost nothing, and, he thought, who would remain in the background. Hahn took Meitner on. In 1908, when Prussia changed its policy and admitted women to universities, for the first time Meitner was allowed to venture upstairs and no longer did she have to walk eight blocks to a hotel for a bathroom. Within a decade Meitner would be in charge of a division of the new Kaiser Wilhelm Institute, and Otto Hahn would defer to her superior knowledge of radioactivity.

In France there were two other laboratories that measured the strength and content of radium salts and various radioactive isotopes. Neither functioned efficiently or accurately. Marie Curie stepped into the breach. Curie's authentication service began in 1911. A numbered certificate was given by the laboratory but could not be used for advertising. Marie wrote, "For measurements dealing with new questions, only my laboratory . . . is in a position to solve the problems which arise." The Curie laboratory was the primary authority on the metrology of radioactivity. Its certificate was valuable and unimpeachable. If a radium preparation contained mesothorium, it became difficult to determine how much gamma radiation was attributable to each substance. (Mesothorium is either of two decay products of thorium: mesothorium I is an isotope of radium, and mesothorium II is an isotope of actinium.) This was hard work and, though at the time it was overlooked, deadly work. Most of all, it was profitable. Jacques Danne, one of Pierre's assistants, now worked for Marie, but seldom put in his required time at the factory. In 1909, Madame Curie heard that Danne planned to open his own laboratory in competition with hers and demanded his resignation:

> Having considered your current situation and the needs of the laboratory, I feel that it is no longer possible to do the things I need done and I ask you to resign from your post as an assistant immediately. I need to be assisted in my work and have someone who is always there and completely at my disposal.

Danne left and with his brother Gaston did open his own laboratory. When pitchblende mines finally were discovered in South Terras in Cornwall, he negotiated a lower price than that of the Curie laboratory to measure radioactivity and established alliances with an extraction factory, as well as a workshop to fabricate instruments, a physics and radiotherapy research laboratory, and a training center, all patterned on her laboratory. With a group of investors, the Danne brothers formed the Industrial Radium Society (Société Industrielle du Radium) at Gif-sur-Yvette, France, and issued a prospectus lifted word for word from an independent report written at the Curie laboratory. Marie, though angry, did nothing about this. She agreed with Rutherford, who observed, "I think it is a great pity when radioactive people start squabbling on the financial side of radioactivity."

Various strengths of radium salts were being manufactured at factories throughout the world, and often they proved to be much weaker than advertised. For these valuable substances, it was essential there be no errors in the strength of the quantities sold. The need for a standard was pressing, and Marie had developed her own based on measuring how much weaker or stronger the radioactivity was compared with that of pure radium. In 1908, the chemist Bertram Boltwood at Yale University's new Sloane Physics Laboratory in New Haven, wrote, asking Marie to compare the standard she was using for her radium with his own. She refused. Boltwood wrote Rutherford, "The Madame is not at all desirous of having any such comparison carried out, the reason, I suspect, being her constitutional unwillingness to do anything that might directly or indirectly assist any worker in radioactivity outside her own laboratory." Her refusal so angered Boltwood that on Curie's subsequent trip to America he used the power of his position to prevent Yale from granting her an honorary doctorate. Boltwood had also blocked Einstein from receiving this honor because he was a Jew.

When the International Congress of Radiology and Electricity convened in Brussels in September of 1910, Marie was still seeking to keep radium under her control. She disingenuously told the gathering that she and Debierne had obtained a precise standard, but she was keeping it sequestered in her laboratory to check the gamma radiation level. In the cat and mouse negotiation that followed, Rutherford suggested that the International Radium Standard Committee buy her specimen. Marie replied that "for sentimental reasons . . . I wish to retain the standard *chez moi*." She would not budge from this position. Fellow scientists enlisted Rutherford to "handle . . . this prickly woman."

At a second session, it was diplomatically put forth that the standard unit of measure be named a *curie*, which pleased Marie, but along with this they informed her that under no circumstances could the International Standard specimen remain in private hands. Several scientists suggested that they be selected instead of Marie to set the standard. Bertram Boltwood insisted that he was the best choice because a standard could be found not by using Madame Curie's slow measurements of radium's emanations but by a common, though less accurate, method utilizing gamma rays. Sir William Ramsay, a well-known chemist, also volunteered. Rutherford, however, insisted on Madame Curie. For reasons that would soon become apparent, the standardization process was to take over three years.

Soon after the meeting, the rejected Ramsay published his experiment in which radium's emanation, radon gas, when combined with copper, produced lithium. Only Marie Curie had enough strong radium to test Ramsay's results. He had used glass containers in the experiment. Marie used platinum containers to prove that Ramsay's use of glass containers caused the lithium to infiltrate the mix. Ramsay's reputation was damaged. Boltwood wrote sarcastically, "I wonder why it hasn't occurred to him [Ramsay] that radium's emanation and kerosene [when combined] form lobster salad." Ramsay did not forget the insult. When asked about Madame Curie he wrote, "All the eminent women scientists have achieved their best work when collaborating with a male colleague."

Marie Curie was providing invaluable work, but throughout her career sexism and jealousy would cause her fellow scientists to diminish her accomplishments. When she began work on the radium standard, William Thomson, Lord Kelvin, at the age of eighty-two, in an act that one doubts would have happened had she been a male scientist, wrote a letter to the *London Times* stating that Madame Curie's radium was not an element at all but a helium compound. He, too, had an axe to grind: Lord Kelvin was famous for his studies that had set the age of Earth at 20 million to 50 million years. Marie's discovery of radioactivity, followed by her substantiation of Rutherford's transmutation theory, had put the age of Earth at twice or more Lord Kelvin's figure. The "old boy" was not pleased. He did not write his attack on Madame Curie for a scientific publication, which would have been appropriate, but for a major newspaper aimed at a wide readership. Soon the debate was joined by scientific publications. To Marie's astonishment several well-known physicists agreed with Lord Kelvin. Marie said and wrote nothing. For her there was only one way: Characteristically she set out to provide more scientific evidence of her discovery no matter how long it took or how laborious the task. First, she conducted an even more specific study of the atomic mass of radium and published a result of 226.45 plus or minus 0.5.

(The current atomic weight is 226.025.) Then, with a determination that was unique, she set out to put the question to rest once and for all by creating radium as a pure metal for chemists to see and touch.

Working with Debierne, she spent three years on a task as arduous as her original quest for pure radium. After all this time, she produced a tiny square of shiny white metal with a melting point of 700 degrees that darkened almost immediately when exposed to air. This experiment, in Marie's words, "has never been repeated . . . because it involves a serious danger of loss of radium which can be avoided only with utmost care. At last I saw the mysterious white metal, but I could not keep it in this state, for it was required for further experiments."

With the publication of her two-volume *Treatise on Radioactivity* in 1910, an exhaustive study which lucidly describes all that was then known about radioactivity, several scientists reacted with barely concealed jealousy. Rutherford, while admitting that he wished he had written the book and that "Madame Curie has collected a great deal of useful information," simultaneously wrote that she had included "too much information" and concluded, "The poor woman has labored tremendously and her volumes will be useful for a year or two to save the researcher from hunting up his own literature: a saving which I think is not altogether advantageous." Rutherford's contemporary and biographer, A. S. Eve, wrote him on reading the treatise that it had reminded him of the earlier battle to prove radioactivity emanated from inside the atom. "It is astonishing how slow [the Curies were] to accept your conclusions. . . . I believe if it had not been for you the whole subject would have been a grotesque muddle to this day." As a pupil of Marie's observed, she had written this book not only as "a passionate researcher . . . [but] to prove to those who kept insinuating it, that she was not simply Pierre Curie's assistant in their common work." Today this extraordinary work is still regarded as the most accurate account of the early history of radioactivity.

With rich sources of radioactive ore being discovered around the world, the Curie laboratories had increased in importance but not in scale. The de Lisle factory laboratory was fully occupied with equipment overflowing into the halls. On the rue Cuvier, Curie's staff was also working in cramped quarters, without government subsidies even for equipment. Austria, seeing an opportunity, offered to create a state-of-the-art laboratory for Madame Curie. Although there was little chance that she would leave France, the offer acted as a goad for negotiations to begin with the Pasteur Institute to create a radium institute for her. The new facility would be part of the institute and consist of the Curie and Pasteur Pavilions with a garden in between. Madame Curie was now at the height of her career, and her laboratory was to be the one that Marie and Pierre had dreamed of. But Pierre was dead, and throughout history, once an icon has been created, there is a societal compulsion to destroy it. Marie Curie's downfall was to be as fierce as a Greek tragedy.

CHAPTER 16

Honor and Dishonor

On a warm April night in 1910, Marie dropped in at the Borels, who were having an informal dinner with the Perrins. Instead of the severe black dress she had worn since Pierre's death she now wore a fashionable white gown with a single pink rose pinned to her waist. She seemed transformed; the hard scowl replaced by a subtle relaxation. "What happened to her?" Jean Perrin asked Marguerite Borel the next morning.

Although she professed to have no "social life" and her daughter Eve later wrote that almost no one knew her intimately, Marie did have a small coterie of loyal friends consisting mostly of those who understood her work: There were the dedicated André Debierne and Jean Perrin, an expert in cathode rays, the disintegration of radium, and the composition of heat and light. Jean's wife, Henriette, like Marie's sister Bronya, was a calming presence and addressed Marie with the intimate *tu*. Then there were the Borels: Émile, who had been named dean of the École Normale Supérieure, and Marguerite, the daughter of Paul Appell, dean of the School of Sciences at the Sorbonne. Hertha Ayrton, who was a well-known scientist and pioneer in England's women's rights movement, also was a close friend. Although Hertha lived in London, distance did not impede their friendship. Both were outsiders: Marie, Polish; Hertha, Jewish. The author George Eliot (Mary Ann Evans, who chose to write under a male pseudonym) had helped subsidize Hertha's education and had based the character of Mirah, the talented Jewish outcast in *Daniel Deronda*, on her. All of these friends were to be involved in what was soon to be called the "Great Scandal."

What had happened was Paul Langevin. Pierre's former student had long been a dear friend of the Curies and he was Pierre's chosen successor at the EPCI. Five years younger than Marie, he was a tall man with military bearing, penetrating eyes, a severe brush haircut, and a fashionable handlebar mustache. Langevin was both a physicist and a brilliant mathematician. In 1906, Langevin had reached the conclusion that $E = mc^2$ (energy equals mass times the speed of light squared), only to find that a fellow scientist named Einstein had already published this discovery.

Marie Curie wrote to Henriette Perrin that she "greatly appreciated [Langevin's] wonderful intelligence." He helped her prepare her course lectures at the Sorbonne and refined her presentation. She found him a sympathetic friend who was soon asking her for advice on what he termed his "disastrous mistake of a marriage" to Jeanne Desfosses, the daughter of a working-class ceramicist, who he felt held him back from great discoveries through her violent nature and constant demands for money. Langevin wrote that he was drawn to Marie "as to a light . . . and I began to seek from her a little of the tenderness which I missed at home." Jeanne Desfosses Langevin welcomed Marie into their household, where Marie met the Langevins' four children. In the spring of 1910, Jeanne complained to Marie about Paul's cruelty toward her and Marie chastised him. In return, he showed Marie a half-healed gash where Jeanne had broken a bottle over his head.

Most of what we know of the Curie-Langevin relationship comes from friends' accounts, and most significantly from letters Marie wrote to Paul which a detective in his wife's employ purloined from the desk at the small apartment near the Sorbonne that Langevin had rented. By July of 1910, these letters suggest that Marie and Paul had become lovers. Here was a friend, soul mate, and potential partner in science who might replace Pierre. It would be a second chance for Marie to repeat the best days she had known. With this fervent wish she wrote him,

> It would be so good to gain the freedom to see each other as much as our various occupations permit, to work together, to walk or to travel together, when conditions lend themselves. There are very deep affinities between us which only need a favorable life situation to develop. . . . The instinct which led us to each other was very powerful. . . . What couldn't come out of this feeling? . . . I believe that we could derive everything from it: good work in common, a good solid friendship, courage for life and even beautiful children of love in the most beautiful meaning of the word.

Although she had tolerated her husband's past infidelities, Jeanne Langevin, upon first suspecting his relationship with the famous Madame Curie, flew into a rage threatening to kill Marie. Perrin momentarily calmed Jeanne, but she and her sister waited in a dark street near Marie's apartment. As Marie walked by, Jeanne accosted her and ordered her to leave France immediately or die. Afraid to return to her house, Marie fled to the Perrins. Jean Perrin noted, "This illustrious woman had been reduced to wandering like a beast being tracked." Paul Langevin advised Marie that his wife was entirely capable of murder and advised her to leave France. She refused. Finally, it was decided that temporarily the two would no longer see each other. But when Langevin and Curie left Paris for the International Congress of Radiology and Electricity, Jeanne Langevin told her sister that the trip was only a subterfuge to hide their affair. She renewed her threats against Marie and threatened to expose them. When Marie arrived at the conference, Rutherford was the first to notice her condition. He wrote, "Madame Curie looked very worn and tired and much older than her age. She works much too hard for her health. Altogether she is a very pathetic figure." Stefan Meyer, who had developed his own radium standard, was more cynical and told Rutherford that the very visible attacks of nerves and exhaustion that caused her to leave committee meetings only occurred when the discussion displeased her.

After the conference Marie and Paul returned to Paris, and then she joined her children in l'Arcouëst, on the northern coast of Brittany, a preferred summer gathering place for scientists and professors (so much so it was nicknamed "Fort Science"). The Borels and Perrins were in residence. Marguerite Borel had become Marie's close friend and confidante. One night Marie grabbed Marguerite's hands and poured out her fear that though she would walk through fire for Paul

Langevin he might yield to Jeanne's pressure, desert science for a more lucrative profession, or sink into despair. "You and I are tough. . . . He is weak." In spite of this, just as she had with Casimir Zorawski, she deluded herself that they would find a way to be together.

Marie avowed her love for Paul and dramatized the fact that she was risking her reputation for his sake and might even commit suicide if things did not work out: "Think of that, my Paul, when you feel too invaded by fear of wronging your children; they will never risk as much as my poor little girls, who could become orphans between one day and the next if we don't arrive at a stable solution." In what can only be interpreted as a fit of jealousy, Marie cautioned Langevin that if he resumed sexual relations with his wife and if she had another child they would both be "judged severely by all those, alas already numerous, who know. If that should happen it would mean a definite separation between us. . . . I can risk my life and my position for you, but I could not accept this dishonor. . . . If your wife understands this, she would use this method right away."

This was followed by several letters to Paul, instructing him, in a mixture of pragmatic cruelty and passion that demonstrated her insensitivity, on how he might rid himself of his wife. "Don't let yourself be touched by a crisis of crying and tears. Think of the saying about the crocodile who cries because he has not eaten his prey, the tears of your wife are of this kind." She pleads with Langevin, "When I know that you are with her, my nights are atrocious. I can't sleep, I manage with great difficulty to sleep two or three hours; I wake up with a sensation of fever and I can't work. Do what you can and be done with it. . . . We can't go on living in our current state." Marie, who had to be cajoled into marrying the placid Pierre, was now aflame. A letter ended, "My Paul, I embrace you with all my tenderness. . . . I will try to return to work even though it is difficult, when the nervous system is so strongly stirred up."

Langevin, however, seems to have been ambivalent. Once before he had separated from his tempestuous wife only to beg her to return. Langevin did not leave his wife nor did he stop seeing Marie. The usually quiet André Debierne had a loud argument with Paul Langevin, blaming him for Marie's increasing bad health and emotional outbursts. She seemed distracted at work and paid little attention to her daughters.

The situation was made worse by a series of disappointments that struck Marie one after another: At the urging of friends and perhaps to make Langevin proud, she announced her candidacy for a Chair in Physics at the Academy of Sciences, the most powerful scientific body in France. Members read their papers, met for symposia, and gave large grants for scientific study. The other applicants for this chair were weak, save for Édouard Branly, an inventor who was instrumental in helping Marconi develop the wireless telegraph.

For this elite male organization, Marie Curie's action came as a bombshell that resulted in negative comment not only from men but from women who found her a threat to their femininity. "Science is useless to women," wrote the influential writer Julia Daudet, and Madame Marthe Régnier, the famous actress, wrote in *Le Figaro*, "One must not try to make woman the equal of man." The right-wing tabloid newspapers loosed a barrage of criticism, bringing up Marie's Polish origins and antiwar statements. It was a harbinger of what was to come.

On Monday, January 24, 1911, members of the Academy of Sciences gathered for a vote. President Armand Gautier announced that everyone was welcome to enter the chamber except women. On a second ballot Édouard Branly received thirty votes, Marie Curie twenty-eight. Her loss engendered sympathetic letters from many scientists, but the fact remained that Madame Curie, France's most famous scientist and Nobel Prize winner, could not present her own papers at the academy nor did she ever try to do so, writing instead for scientific journals such as *Comptes rendus.*

By the spring of 1911, Marie and Paul, unable to separate, were once again secretly meeting in Langevin's rented Paris apartment, but Marie was worried that Jeanne was having her husband followed and even that his eldest son might be spying on them. It was at Easter that the intimate letters stored in a desk drawer disappeared. A week later Jeanne Langevin's brother-in-law paid a visit to Madame Curie and told her that these letters were now in Madame Langevin's possession and she was prepared to make them public. Paul Langevin, in a rage over the stolen letters, left home for two weeks but returned. On July 26, after another fight with his wife, Paul left again, and Jeanne filed charges of abandonment.

Marie, worn down and frightened, sent Irène and Eve to visit the Dluskis' in Poland. At the end of the summer she joined them, then left for Brussels and the 1911 Solvay Conference. These conferences, which attracted the greatest scientific minds, were underwritten by Ernest Solvay, an affluent Brussels chemist and philanthropist who had developed a new process for manufacturing sodium carbonate. Once again Paul Langevin was there, as were Jean Perrin, Albert Einstein, H. A. Lorentz, Max Planck, and Ernest Rutherford among

The Solvay Conference, 1911. Marie Curie and twenty-three male scientists. Poincaré on Curie's left, Perrin on her right. Standing: Rutherford behind Curie; Einstein and Langevin, far right.

others. In a thrilling moment while at the conference, Madame Curie received a telegram from the Nobel committee announcing that she was the sole winner of a second Nobel Prize, this time in chemistry. In the official letter that followed, she was commended for "producing sufficiently pure samples of polonium and radium to establish their atomic weight, facts confirmed by other scientists, and for her feat of producing radium as pure metal." Almost simultaneously a second telegram informed her that Jeanne Langevin had released her letters to the press. Marie left abruptly but not before hastily writing a note to Rutherford saying that she appreciated that he had appointed her to create the radium standard, and was touched by all the kind attention he had shown her during the conference. She explained that she had hoped to shake his hand before leaving but was ill and couldn't stay.

Marie returned to France and to venomous publicity. The vindictive Boltwood declared, "She is exactly what I always thought she was, a detestable idiot!" Marie Curie's house was surrounded by people who threw stones at her windows. She fled with her children to the Borels. The press printed her explicit instructions to Paul Langevin on how to get rid of his wife and accused her of being a home wrecker, a dissolute woman, a Polish temptress, a Jew.

The right-wing tabloid press of the day had helped bring Marie Curie fame, and put the Nobel Prize, which had been

How dare Madame Curie try to enter the all-male Academy of Sciences? Prejudice against her was building and would soon explode.

little noticed in the field of science, on the map. Now not only did the press topple an idol, but in so doing it catered to both factions of a divided public. For half a century, after France's humiliating defeat in the Franco-Prussian War, followed by the brutal extermination of the Paris Commune, there had been political discord in France. The sole issue on which the majority agreed was a desire for revenge against the foreign invasion and the dishonor which France had endured. Anti-Semitism, chauvinism, and xenophobia became the keynotes of the powerful right. Gabriel Lippmann, who refused to condemn Madame Curie, was called "the Jew of color photography," and Jean Perrin, who defended her, was branded "a fanatic Dreyfusard." (In a burst of anti-Semitism in 1894, this same right-wing press had hastened the conviction of the Jewish captain Alfred Dreyfus, falsely accused of being a spy.)

Irène was at school when a friend pointed to a headline about the affair in *L'Oeuvre*. She read the story and burst into tears. The ever-faithful Debierne arrived and took her to the Borels. In the Borels' guest room the frightened Marie awaited her children. Eve arrived and told Marguerite Borel, "Mé is sad, a little sick," and then from what must have been her own desire, "Mé needs cuddling." The stoic Irène stayed close to her adored mother, who huddled in a corner and, in a rare gesture of affection, stroked Irène's hair. The Perrins arrived and volunteered to take Irène to their home. She refused insisting, "I can't leave Mé." Finally she was persuaded to leave. Although all attention was focused on Marie, it was as she had predicted, her daughters too were suffering.

Paul Appell, who with Debierne had delivered Pierre's corpse to Marie, turned on her as did many former friends. He arranged for a group of professors at the Sorbonne to demand that Madame Curie leave France. When Appell heard that his own daughter had taken Marie to live with her, he called her to his apartment. Marguerite Borel arrived to find her father in a rage. "Why mix in this affair which doesn't concern you?" he demanded. He announced that the next afternoon he was planning to see Madame Curie and demand that she leave the country. He had arranged a chair for her in Poland. "Her situation is impossible in Paris. . . . I can't hold back the sea which is drowning her."

By her own account, Marguerite, who never in her life had dared oppose her father, stood before him trembling. Then she drew herself up and answered, "If you yield to this idiotic nationalistic movement, if you insist that Madame Curie leave France . . . I swear that I will never see you again in my entire life." Appell, who had been putting on his shoes, hurled one across the room. Beneath his anger was the fear that his daughter would be "swept away" by this scandal, but he gave in and agreed to postpone his decision.

Marguerite observed that none of this would be happening if Madame Curie were a man. Indeed no one had asked Paul Langevin to leave the country or condemned him. It was well known that Albert Einstein had sired an illegitimate daughter whom he might have put up for adoption, but in any case she was never seen again. In 1911 every day, barring weekends, an average of thirty-nine declarations of adultery were recorded in Parisian courts. Twenty-four births in every hundred were illegitimate. Marie's sin above all was that she was not just a mistress but an emancipated woman when such women were regarded by both sexes as a threat. With the abortion rate at an all time high, newspapers had formed a coalition to eliminate advertisements for midwives offering "discreet services." And worse, here were letters that showed a passionate woman: A respectable woman was supposed to endure sex, not relish it. For men the pleasures of sex were to be found outside the marriage bed. The press continued their attack like vultures feeding on carrion. Bronya and Casimir Dluski rushed to Marie's side and Jacques Curie defended her. The whole affair had by now taken on an *opéra bouffe* tone. The right-wing journalist Gustave Téry wrote that Langevin was "a boor and a coward." Langevin challenged him to a duel. Duels were illegal but frequent. After elaborate preparations Téry refused to fire his pistol, saying that he did not wish to deprive France of one of its greatest minds. Langevin too never raised his pistol. "I am not an assassin," he said. That ended that.

Madame Langevin, having succeeded in wounding Marie more than she had hoped, finally signed a separation agreement that did not mention her. Three years thereafter the Langevins reconciled and Paul took another mistress, an anonymous secretary. Several years later, after having an illegitimate child with one of his former students, he asked Marie to find her a position in her laboratory, and she did.

CHAPTER 17

"She Is Very Obstinate"

Paul Langevin weathered the storm. Marie Curie did
not. Shortly after the scandal broke upon this fragile
woman, a member of the Nobel Committee wrote her
on behalf of the committee asking her to refrain from coming to Sweden to accept her prize. He cited her published love
letters and "the ridiculous duel of M. Langevin" and added in
a stinging rebuke, "If the Academy had believed the letters . . .
might be authentic it would not, in all probability, have given
you the Prize. . . ."

This judgment added immensely to her pain, but she wrote
back addressing an issue that provokes controversy even now:

> You suggest to me . . . that the Academy of Stockholm, if it
> had been forewarned, would probably have decided not to
> give me the Prize, unless I could publicly explain the attacks
> of which I have been the object. . . . I must therefore act
> according to my convictions. . . . The action that you advise
> would appear to be a grave error on my part. In fact the

Prize has been awarded for discovery of Radium and Polonium. I believe that there is no connection between my scientific work and the facts of private life. . . . I cannot accept the idea in principle that the appreciation of the value of scientific work should be influenced by libel and slander concerning private life. I am convinced that this opinion is shared by many people.

A seemingly harder, prouder, and more aggressive Madame Curie attended the Nobel ceremony. She was accompanied by her sister Bronya and fourteen-year-old Irène. King Gustaf bestowed the prize and no one spoke of "personal matters." In Madame Curie's acceptance speech she complimented other scientists who worked in the field of radioactivity but firmly established her own credentials. "The history of the discovery and isolation of this substance furnished proof of my hypothesis according to which radioactivity is an atomic property of matter and can provide a method for finding new elements." Taking full credit for her accomplishments, she reminded the committee that "isolating radium as a pure salt was undertaken by me alone." When she had finished, there could be no doubt of who had made these great contributions to science.

Marie then returned to Paris. Nineteen days later she was rushed to the hospital with what was said to be a kidney ailment. Some doctors diagnosed old lesions pressing on her kidney. Others thought that perhaps she suffered from asymptomatic tuberculosis. What was not said was that she had experienced a total nervous breakdown and had fallen into the deepest, darkest depression of her life, more enduring than all the episodes that had come before. Later she told her daughter Eve that this time she wanted to kill herself and indeed some of her letters indicate that she planned to commit suicide. She refused to eat and her weight dropped from 123 pounds to 103. From the hospital she was sent by ambulance to be cared for by the Sisters of the Family of Saint Marie, which dealt with both medical and psychiatric ailments. In February she underwent a kidney operation. She was convinced that she was about to die and instructed Debierne and Georges Gouy how to dispose of her precious radium.

While Marie lay in a perpetually darkened bedroom, Ernest Rutherford transformed the entire concept of the atom. In a landmark experiment he aimed alpha particles at a piece of gold foil and observed that a few stray particles bounced back from the foil. This was, in Rutherford's words, "almost as incredible as if you fired a 15-inch shell at a piece of tissue paper and it came back at you." J. J. Thomson's theory, which was the best any scientist had come up with, held that the interior of the atom consisted only of electrons scattered throughout as in a "plum pudding." But if that were true, then the alpha particles traveling at great speed would have moved right through. Rutherford postulated that the atom consisted largely of empty space, but with a dense central core, which he named the "nucleus." When an alpha particle struck the nucleus, it bounced back.

Shortly before her collapse, Marie had decided to leave the house in Sceaux, which had become a tourist attraction, and in her absence her daughters moved to a new apartment on the Île St. Louis in Paris in the care of a Polish governess. The apartment was barely furnished. They did not see their mother again for almost a year, when she visited briefly with them at a house in Brunay which she had rented under the pseudonym of Bronya Dluski. In July, after taking water treatments for her nerves, she moved once again, this time using the name, Madame Sklodowska.

André Debierne tried to fill the gap, forwarding her mail and informing the family of her whereabouts. Debierne looked out for Irène and Eve, writing Marie that little Eve seemed unduly nervous in the absence of her mother. At Christmas 1912, he noted that Eve was particularly sick. Marie roused herself and spent the holiday in Lausanne with her daughters while Eve recovered. Then Marie left. In the spring of 1913, Eve was treated for worms. Once again her mother returned, and Eve recovered.

Debierne also kept a close eye on Irène, who, in the absence of the mother she worshipped, tried to become more independent each day. She showed no emotion, though later she wrote of her suffering. Marie had adopted a series of pseudonyms, but when Irène was told not to write to her mother under her true name, she felt humiliated. Irène took great pride in the maligned Curie name and vowed always to keep it. She became obsessed with finding out as much as she could about her father's family. In the spring of 1912, she wrote her absent mother for permission to visit her uncle Jacques Curie, who so resembled her dead father.

Marie Curie's gradual recovery should be credited largely to Hertha Ayrton, who offered Marie and her daughters sanctuary in England. Using her mother's name, Marie met Hertha at an old mill-house she had rented for the summer at Highcliffe-on-Sea in Hampshire. In addition to being a scientist of note, Ayrton was a superb nurse and a crusader for women's rights. Two years previously the National Suffrage League had been formed, and working women forced through a law permitting them to keep their salaries instead of turning them over to their husbands. The English suffragettes were extremely militant. Hertha changed the way Irène thought about these women. Earlier that year Irène had written her mother, "I've noticed that every day or almost every day an English minister just misses being killed by the English suffragettes, but it seems to me that that isn't a very brilliant way for the suffragettes to prove that they are capable of voting." Now she was more sympathetic.

The British way of coping with women who demonstrated for their rights was to arrest them, but when they went on hunger strikes the authorities did not force-feed them as they had in the past, but left them without food and released them only when they were near starvation. When Christabel Pankhurst, a leader of the movement, was released, she was so weak that she had to be carried out of jail on a stretcher. Hertha nursed Pankhurst and many other suffragettes back to health. She applied her skills to Marie, who little by little began to function but never again would regain the dynamic force she had previously demonstrated. At the end of the summer Marie was too depressed to attend the 1912 Solvay Conference, which was dedicated to finalizing the radium standard.

In her debilitated state, it fell on Debierne's shoulders to continue the work of the research and the factory laboratories, overseeing the manufacture of radium salts that were now providing Marie with a reasonable income. In her absence the squabbling about the standard began all over again. Rutherford, who was in charge, delegated Debierne to act for Madame Curie, finding him to be "a very sensible fellow." From her sickbed, she asked to postpone delivery of the radium standard sample. Once again, she asserted that this original sample should be kept in her laboratory. Rutherford told Debierne, "She is very obstinate," and refused, maintaining that a sample in the name of the International Standard could not be held by "herself as an individual." Debierne suggested a compromise: a duplicate sample could remain in the Curie laboratory, but the original would be moved to the Bureau of Weights and Standards in Sèvres. Madame Curie finally agreed but demanded payment or replacement of the radium used in both samples. Rutherford gave in. A year later in 1913, she had recovered sufficiently to travel to Sèvres carrying a glass tube, which she had sealed herself, containing 21.99 milligrams of radium chloride. The International Radium Standard Committee, as promised, agreed to call the standard based on her sample a "curie."

In July of 1914, the construction of the Curie laboratory at the Pasteur Institute was almost complete. Marie oversaw the planting of trees and a rose garden between the two pavilions. The Curie Institute had finally become a reality, but there was to be yet another postponement. On August 3 Germany declared war on France. As Marie emerged from her mental and physical collapse, World War I began.

CHAPTER 18

"All My Strength"

September 3, 1914: At Paimpol station, the forty-seven-year-old prematurely grey woman, dressed inconspicuously in a heavy black alpaca coat, maneuvered awkwardly through the scurrying crowds of luggage-laden men, women, and children. The German army had entered France and was advancing toward Paris. Two days earlier the president of France, Raymond Poincaré, had relocated the government to Bordeaux. Madame Curie's suitcase contained tubes of radium bromide encased in lead, representing all the radium in France. Her mission was to insure that the advancing German army would never capture what officials deemed "a national treasure of inestimable value."

The trip to Bordeaux took ten hours as the train stopped frequently at railroad crossings. From the windows Marie could see that the roads were clogged with vehicles of all kinds as people fled. She was embarrassed to be seen on this train lest someone think she did not have the courage to remain in Paris in the face of danger. It was night when the train finally pulled into the Bordeaux station. Marie waited for the crowd to disperse and then stood on the platform, the suitcase at her feet. A government representative finally appeared. All the hotels were filled with fleeing Parisians, so he drove her to a shabby private home where she spent the night in a cramped room, the suitcase at the foot of her bed.

A puzzling question remains. How could Marie Curie carry a suitcase loaded with lead? Hélène Langevin-Joliot had never been asked this question and it intrigued her. She speculated:

> About the trip to Bordeaux: I found no description of the protecting box. On the other hand, we have the lead cylinder used to transport the one gram of radium received from the American women after the war. From the geometry of the case, my estimation of the weight is somewhere between 40 and 50 kg. This is not enough for the presently required protection. It may well be that Marie used even a smaller amount of lead for the urgent trip to Bordeaux, for example 5 cm to 7 cm of lead around the radium vials, which means a suitcase of about 12 kg to 31 kg. This is heavy, but not impossible to lift, even for a woman. Most probably, somebody from the Institute took the suitcase to the train for Marie in Paris. Maybe she was waiting for somebody to help in Bordeaux? Looking at a woman with a heavy suitcase, somebody in the train may well have helped her to transport it from the train to the station platform.

The following morning, Marie was driven to the University of Bordeaux, where the radium was stored in a vault, and then she was taken directly to the railroad station, where she boarded a train full of soldiers called to duty in Paris.

Fearful of a German invasion of Paris, few civilians were on the returning train. Marie, who had not had any food in a day and a half, was grateful when one of the young soldiers shared his sandwich with her. Then he asked, "Are you not Madame Curie?" To which she whispered her standard answer, "No. It is a mistake." In the following week the woefully unprepared French rushed young men to the Front in taxicabs, and any other available vehicles, where they hastily teamed up with British forces to repel the invading Germans. On September 5, 1914, the Battle of the Marne began.

Friends became enemies. Georges Gouy and Émile Borel were called to the Front. Paul Langevin enlisted as an army sergeant. The organic chemistry division of the Kaiser Wilhelm Institute was converted to "a military installation with a staff of over two thousand, including 150 university professors." Otto Hahn devised mustard, chlorine, and other deadly gases, and, although Lise Meitner worked as a volunteer technician, several sources credited her with assisting Hahn in this work. The French counterattacked with gases of their own. Hertha Ayrton invented a fan to remove the poisonous gases from the trenches. Hans Geiger was twice wounded in Austria. Henry G. J. Moseley, the most promising of the young British scientists, was killed at Gallipoli in 1915.

When Marie had returned to Paris, the city was nearly deserted. All work had stopped at the Curie Institute. The prediction had been that this would be a swift war, but soon rumors began to drift back to Paris of the terrible carnage on the Western Front. Marie hated war. "Only through peaceful means can we achieve an ideal society," she wrote to Hertha Ayrton. "It is hard to think that after so many centuries of development, the human race still doesn't know how to resolve difficulties in any way except by violence." With men gone, women stepped into the breach: Marguerite Borel ran a hospital; Henriette Perrin became a volunteer nurse. Even though she had been excoriated, Marie Curie, too, "resolved to put all my strength at the service of my adopted country." Maimed soldiers were returning to Paris, limbs hacked off and bodies destroyed by probing because no X-ray equipment or technicians were available at field hospitals. Within a few weeks, she commandeered unused X-ray equipment from laboratories and the offices of doctors whose occupants had gone to war. At first, this equipment was placed in Parisian military hospitals. Then, in a moment of inspiration, Marie devised the idea of "mobile X-ray units" which could be used in battle-front hospitals to diagnose the wounded before treatment. The first two cars were donated by the Union des Femmes de France. The cars had to be small enough to navigate narrow roads and that the equipment must be lightweight.

Each mobile unit contained a small generator that could be hooked up to a car battery when electricity was unavailable on-site. An X-ray tube was installed on a movable stand so that it could easily be wheeled to the crucial area. There was a folding table for the patient, photographic plates, a screen, heavy curtains to exclude light, and ampules filled with radon (emitted as the gaseous decay product of radium). For protection there were cotton gloves and a lead-filled apron. It was the perfect marriage of technology and practicality.

Eve and Irène had scarcely seen their mother in two years. Afraid for their safety were Paris to be occupied by the Germans, Marie sent them to l'Arcouëst with their Polish governess and a maid. Irène, at almost sixteen, was assigned the job of looking after Eve, then nine, and wanted none of it. Like her mother, she read books simultaneously in English, German, and Polish and was as devoted to Charles Dickens as her grandfather had been. In her letters to her mother a subtle change was taking place. Irène began to address her as *Ma Chérie* and use the familiar *tu*. When a neighbor overheard the Curie girls speaking Polish to their governess, she accused them of being foreign enemies. Soon after, a man burst into the house and in a drunken rage denounced them as German spies. Irène wrote of this experience to her mother; clearly it traumatized her. She ended the letter, "It pains me to think I am taken for a foreigner. . . . I love France more than anything. I can't keep from crying . . . so I will stop so this letter is legible." Irène kept writing, pleading to return to Paris to help with the war effort. In October, when it seemed apparent that Paris would not fall to the enemy, Marie allowed her daughters to return. Eve was put in a primary school and Irène entered the Sorbonne, where she studied mathematics and physics. In addition to a taxing academic schedule she enrolled in a nursing course.

The mobile X-ray units (now called "Les Petites Curie") were off to a slow start, with bureaucrats forbidding women drivers and technicians to go to the front lines, but Madame Curie prevailed. Dressed in an alpaca coat with a Red Cross armband on the sleeve, she drove to field hospitals at twenty miles an hour. She quickly unloaded the equipment, hooked up a cable to the lightweight generator, covered the windows, unfolded the table, installed the ampoule, and activated the machine.

Marie desperately needed more drivers and technicians and above all someone she could trust. She found all that in Irène,

whom she now considered "my companion and friend." Irène
had replaced Pierre Curie as a collaborator. When Irène was
seventeen, her mother took her out to the battlefields and
after a few months left her alone in charge of a field radiolog-
ical facility in Hoogstade, Belgium, where she worked within
earshot of gun and cannon fire. Soldiers were carried into the
makeshift medical tent, some dead, some missing limbs, some
with shattered bones and shrapnel wounds. By the beginning
of November, French casualties numbered 310,000 dead and
another 300,000 wounded.

Irène, alone and unassisted, X-rayed the wounded, those
young men who in another time might have been her dancing partners or given her a first kiss. After completing the X-
ray process, primly, deliberately, Irène made a geometric
computation that revealed the exact location of bullets and
shrapnel. She then directed the surgeons exactly where to
probe. The surgeon in charge of the hospital had expected
that the X-ray alone would instantly reveal a location and the
calculation stymied him, as did the young woman who was
telling him his business. He probed the wounded at random
and unmercifully until finally he followed Irene's directions
with success.

Irène spent her eighteenth birthday training nurses to take
her place when she moved on to another battlefield position.
She wrote proudly to her mother that on that birthday she
had located four large shell fragments that were successfully
extracted from a soldier's hand. To celebrate she went to a
soccer match and a concert and then slept in her muddy tent.
She ended, "I spent my birthday admirably . . . except that you
weren't there, Ma Chérie."

After Hoogstade Irène moved on to Amiens. She taught

herself to repair equipment, trained nurses, lived like a soldier. By 1916 she returned to Paris to teach a training course
for women X-ray technicians at the new Edith Cavell Hospital, named for the English nurse who had been executed by
the Germans the previous year. These technicians, eventually
about one hundred fifty women strong, were dispatched to
field X-ray posts. In addition to her taxing teaching schedule,
Irène enrolled at the Sorbonne and graduated with honors in
mathematics, physics, and chemistry.

On land, the war on the Western Front was at a stalemate.
At sea, German submarines were proving lethal. By 1917, two
thousand six hundred and seventeen ships had been sunk by
German submarines. Britain depended on supplies by sea,
and a worried Winston Churchill wrote, "The submarine war
will have starved us into unconditional surrender. Our success has been hanging by a thread, a very slender thread, and
one that is in great danger."

Paul Langevin, inspired by Pierre Curie's work in piezoelectricity, thought that if he could pick up sound waves
beyond the range of the human ear, undersea maneuvers
might be detected. Rutherford had suggested using a con-
denser or a carbon microphone to catch this sound, but he
was lagging behind Langevin in his experiments. Marie, who
had remained a friend of Paul Langevin's, since society would
allow her nothing more, lent him a section of Pierre's piezoelectric quartz that was mounted on the wall of her office.
Langevin removed the mount, recharged the quartz, and used
it to make a microphone that could capture ultrasonic vibra-
tions like those that might be transmitted through sea water
from a submarine. By the middle of 1917, his primitive apparatus, which could detect sound with a wavelength as short as

one ten-billionth of a millimeter, was put into service. Sonar was born, and became a great aid in Britain's battle at sea.

On November 11, 1918, Marie and Irène were working side by side with another scientist at their new institute when the sound of cheering, music, church bells ringing, and cannon fire floated in through the windows. After four years the Great War was over. This bloody conflagration left ten million dead and twenty-one million wounded. In France, of those mobilized, 1,333,000 (or 16 percent) of the country's precious youth had perished.

In the four years of war, thousands of women had worked in French factories, served as nurses and technicians, run schools, hospitals, farms and transportation, but when it was all over, once again they were relegated to prewar traditional roles. The movement toward women's rights had gained momentum, but it would be another twenty-seven years before women were enfranchised in France.

During the Russian Revolution, Poland, through the intervention of Lenin, had briefly come under German rule. After the defeat of the Germans, the Treaty of Versailles in June 1919 stipulated that at last Poland after 123 years was an independent nation. The moment that Marie and the Sklodowska family had waited for had arrived in her lifetime. And, while scientists on both sides had worked on weapons to blind, maim, and kill, Madame Curie had remained faithful to her wish to serve peacefully. During the Great War, over a million X-ray procedures had been performed.

CHAPTER 19

The Making of a Myth

CURIE CURES CANCER! In May of 1921, this headline appeared in newspapers across America. This claim alleviated a fear of death and struck a deep chord in America's collective psyche. Beneath this patently false headline is a photograph that indicates Madame Curie's complicity in the creation of her own myth. Standing next to President Warren Harding, Marie holds aloft a golden key that unlocks a small leather box on a nearby table. Supposedly the box contains one gram of radium, her cancer-curing elixir, purchased with money raised largely by the women of America for her to take to the Curie Institute to continue her work. The real radium was secure in a lead-lined vault, not to be removed until it was delivered to Madame Curie's ship when she was to embark for Paris. How did this come about?

After World War I, a much debilitated Marie went back to work. The French government officially appointed Irène as her mother's assistant. Once again Madame Curie was offered a small state pension and this time she accepted, but it did little to alleviate her professional money worries. Throughout her life the money to work unencumbered had remained a major problem. Other countries provided ample support for science but not France. This outstanding scientist wanted desperately to build the reputation of her new institute and provide the means for her beloved Irène to continue her work after Marie's death. In this regard, she found herself spending a great deal of what she termed "wasted time" fund-raising for the institute, a task she loathed.

Into this troubled atmosphere stepped a dynamic, savvy American: Marie Mattingly Meloney, called Missy, the editor of a well-known publication, the *Delineator*, which concentrated on women's issues. After some difficulty, Missy arranged through Paul Appell to spend "a few minutes" with Madame Curie. Missy told Marie how interested American women were in "her great work." Madame Curie, perhaps sensing an opportunity, replied that American researchers had fifty grams of radium with which to work while her laboratory had "hardly more than a gram."

"But you ought to have all the resources in the world to continue your research," Missy exclaimed, "Someone must undertake this." "Who?" asked the world-weary Marie. By the end of the interview Missy had promised that she would raise a hundred thousand dollars from "the women of America" to buy a gram of radium for Madame Curie's laboratory. Marie did not believe her, but did believe in Missy's sincerity, so she told her that she would have nothing to do with fund-raising, but if Missy succeeded she would consider coming to America "to receive the gift."

Missy Meloney proved to be a skillful fund-raiser. She depicted Madame Curie in the *Delineator* as if she were still impoverished, so poor that she could not afford the radium to continue her work to eradicate cancer, thus implying that Madame Curie used her personal money for this purpose. In fact, Marie was living comfortably in her spacious apartment on the Île St. Louis, owned an apartment building with the Borels and the Perrins, and was buying vacation properties in France. Few Americans understood that she needed more radium for research, not for cancer treatments, or that in 1921 radium was the treatment of last resort, being so expensive that it was used only in parts of the body where X-rays could not reach.

Radioactivity, Marie's great discovery, was hard to explain and had no strong emotional appeal to a lay public, while radium itself was by now well known, appearing in greatly diluted amounts in such popular products as watch dials, beauty aids, quack medicines, and a host of other uses. Missy had simplified matters by describing radium as a sure cure for cancer. Her first editorial for the fund-raising campaign in the *Delineator* was headlined, THAT MILLIONS SHALL NOT DIE. It ended, "The great Curie is getting older, and the world losing, God alone knows, what great secret. And millions are dying of cancer every year."

Missy Meloney organized two fund-raising committees: The women's committee was headed by the founder of the American Society for the Control of Cancer. There was also a prestigious male advisory committee composed of cancer researchers and technicians, but it was the women who contributed donations—large and small—to the Marie Curie Radium Fund. American women had gained the vote the previous year and had begun to assert their equality. The Sunday magazine of the *Daily News*, with the largest circulation in the country, introduced a new comic strip, *Winnie Winkle the Breadwinner*. In Madame Curie, women found a heroine, but also one who needed their help.

Only a year later, Missy Meloney cabled that the goal of $100,000 for a gram of radium had been raised. Marie wrote Missy, "I understand that it will be of the greatest value for my Institute," and with some persuasion she agreed to keep her part of the bargain and receive the gift in person. In France, news of the American fund-raising campaign caused a spate of publicity: Ten years previously she had been almost destroyed by the press, but now Madame Curie was restored to her iconic status. The paper *Je sais tout* arranged a gala celebration at the Paris Opéra. In the audience were the most prominent people in France. Jean Perrin spoke about Madame Curie's great contributions to science and the promise for the future that her discoveries had given their nation. Sarah Bernhardt read an "Ode to Madame Curie" which declared her the sister of the god Prometheus. It was with this send-off that Marie, accompanied by Irène and Eve, boarded the White Star liner *Olympic*.

In New York Missy Meloney handled the press with skill: If reporters wished to interview Madame Curie, she made them promise not to print a word about the Langevin affair. The visit she planned for the ailing Marie would have staggered even a young person in perfect physical condition; eighteen college lectures, seven honorary degrees, a tour of Niagara Falls and the Grand Canyon, dinners, lunches, and a final presentation on the steps of the White House.

When the *Olympic* arrived in New York, the pier was jammed with hundreds of people: Polish patriots, Girl Scouts, nurses and doctors, college students carrying banners, and others who awaited the arrival of this celebrity. Two dozen photographers dashed up the gangway. Flash bulbs popped as an out-of-tune band played the "*Marseillaise*" followed by the "Star-Spangled Banner." Marie froze. She plunked herself down in a deck chair, stared at her feet, and refused to stand up until the photographers left. Irène and Marie were dressed almost identically in unfashionable loose black dresses with felt hats squashed down on their heads. Eve wore a fashionable jacket secured by a gold pin and smiled at the crowd. The press loved Eve and called her "Eve of the radium eyes." They predicted she would marry a prince. Marie could not keep up with the schedule. While her daughters toured the Grand

A publicity-shy Madame Curie arrives in New York City.

Canyon, she stayed behind, too exhausted to travel. At last there was a dinner hosted by President Harding and the following day the symbolic presentation of the gram of radium.

Marie, who by now had become more realistic about how to procure the sums of money she so needed to further her work and that of her daughter, must have been aware that the enhancement of her own myth brought great rewards. In addition to the gram of radium worth $100,000, Madame Curie was given equipment from New York's Sloan Laboratory, $22,000 worth of mesothorium, $7,000 in speaker's fees, an extra $52,000 which Missy had raised, and another $50,000 from the Macmillan Company for the publication rights to a biography of Pierre Curie to be written by Marie. With Missy's advice, Marie insisted that this book be printed in English only and limited to distribution in the United States and Canada. Then, instead of writing the expected biography, she produced a peculiar and propagandistic document. Although this slim book is entitled *Pierre Curie* by Marie Curie, the Introduction, written by Missy Meloney, takes up the first twenty-seven pages and is solely about Madame Curie, depicting her as "a simple woman, working in an inadequate laboratory and living in a simple apartment on the meager pay of a French professor." Missy cements the Curie legend with such prose as the following: "On one morning in the spring of 1898 . . . Madame Curie stepped forth from a crude shack on the outskirts of Paris, with the greatest secret of the century literally in the palm of her hand."

Meloney's Introduction is followed by Marie's brief biography of Pierre Curie, which emphasizes the travails of their scientific life. Marie laments that there was no money or aid in the discovery of radium. One chapter is called "The Strug-

gle for Means to Work . . . The First Assistance from the State,
It Comes Too Late." Mentioning neither the interested scientists who volunteered their time, nor the grants and awards
from scientific institutions and private philanthropists, nor
the invaluable factory assistance in processing the pitchblende
residue, Marie depicts herself exactly as Missy has depicted
her, the conqueror of unconquerable odds.

The last 100 pages of the book consists of Marie's "Autobiographical Notes." Although a translator is credited on the
book's title page, these pages were written in English by
Marie's own hand, and constitute the only autobiography she
ever wrote. They are obviously aimed at the American audience, who knew no more of her than she chose to reveal. Like
Missy's own, Marie's appeal to Americans was highly emotional, calling the substandard laboratory in which she
worked "a miserable old shed":

> Its glass roof did not afford complete shelter against rain;
> the heat was suffocating in summer, and the bitter cold of
> winter was only a little lessened by the iron stove, except in
> its immediate vicinity. There was no question of obtaining
> the needed proper apparatus in common use by chemists.
> We simply had some old pine-wood tables with furnaces
> and gas burners. We had to use the adjoining yard for those
> of our chemical operations that involved producing irri-
> tating gases; even then the gas often filled our shed. With
> this equipment we entered on our exhausting work.

She emphasizes not the tedious fractionations that occupied four years, but the early physical work of the first
months, which she describes in all the grimy detail of a Zola

novel. "Sometimes I had to spend a whole day mixing a boiling mass with a heavy iron rod nearly as large as myself. I would be broken with fatigue at the day's end." In an implicit appeal for donations to the Curie Institute she writes of her discovery of radium. "One year would probably have been enough for the same purpose, if reasonable means had been put at my disposal." In an emotional thank you to "American women" she writes of the gram of radium she received, "I was very thankful to my sisters of America for this genuine proof of their affection." Three years later, when Madame Curie wrote the president of the American Society of Chemical Products requesting polonium for her institute, he replied that it would be an honor to send it to so famous a scientist. In 1929, Madame Curie made a second trip to America, raising the money for another gram of radium, this time for the Warsaw Radium Institute, as well as fifty thousand dollars for her own laboratory. She was lucky. The stock market crashed three days later.

Toward the end of her life, Madame Curie tried to rein in the mythic image she had helped form. She warned that Curietherapy (radiation treatments) provided no sure cure for cancer. Of the "miserable old shed" that she had created in so many minds, she wrote, "It is true that the discovery of radium was made in precarious conditions but the shed which sheltered it seems clouded in the charms of legend." Madame Curie hoped that one day there would be a way to persevere, one better than mythmaking:

Humanity, surely, needs practical men who make the best of their work for the sake of their own interests, without necessarily forgetting the general interest. But it also needs dreamers, for whom the unselfish following of a purpose is so imperative that it becomes impossible for them to devote an important part of their attention to their material interest. No doubt it will be said that these idealists do not deserve riches since they do not have the desire for them. But a well organized society could assure what is for them an essential need: the means of efficient work in a life from which would be excluded the material cares, so that this life could be freely devoted to the service of scientific research.

The Faustian bargain precipitated by the Curies' first Nobel Prize was now complete. Radium, not her other remarkable accomplishments, was what one would most frequently associate with Madame Curie. The myth of radium to eradicate cancer has been equated with Selman Waksman's antibiotic streptomycin, Sir Alexander Fleming's discovery of penicillin, and Jonas Salk's vaccine for poliomyelitis. However, there is a great difference in that the popular use of this element in radiation treatments for cancer and other diseases lasted less than two decades before it was replaced by cobalt and related substances. The medical use of radium is now extremely limited. Oddly enough, not only the public but scholars and even some scientists are unsure about the difference between X-ray and radium treatments. Often, these separate treatments are confused because X-ray and radium applications were both referred to as "radiotherapy" as if they were identical. This confusion may have arisen because Marie Curie, who was identified with radium, also trained X-ray technicians and used X-rays for diagnostic purposes.

In many ways X-rays resemble the gamma rays given off

by radium and other radioactive substances, but there are important differences. Shortly after World War I, Max von Laue of the University of Berlin and his students determined that X-rays were electromagnetic rays identical to, but shorter than, those of visible light. Scientists sometimes describe them as "short-short rays" or "ultra-ultraviolet rays" because their length is significantly shorter than ultraviolet. They are produced by high-velocity electrons bombarding various materials. (Until recently it was thought that X-rays must be produced artificially, but nuclear physicists have now determined that they can also occur naturally in cosmic radiation.)

X-rays began to be used only four months after Röntgen's discovery. They produced clear pictures of the interior of the body and destroyed some diseased tissue. In contrast, the fierce gamma rays from radium were far more penetrating and could reach areas inaccessible to weaker rays. But how to apply this force was not immediately evident. In 1901, Henri Becquerel carried a vial of radioactive barium in his vest pocket for six hours. He felt no pain but fifteen days later a burn appeared on his chest in the exact place and shape of the vial. Becquerel charted his own healing process and told Pierre, "I love it [radium], but I owe it a grudge." Pierre was intrigued by this and exposed his own arm to radium for ten hours to see what medical consequences might emerge. Pierre observed, "one will feel absolutely nothing, but fifteen days afterwards a redness will appear on the epidermis and then a sore which will be very difficult to heal." Even as he warned of the dangers of radium exposure, which had killed laboratory animals, he wrote, "I am happy with my injury. My wife is as pleased as I am. . . . This shouldn't frighten people."

The Curies were convinced that radium would prove to be

of inestimable use in medical treatments, although at first this was not apparent. Pierre donated some of his precious radium salts to the head of dermatology at Saint-Louis Hospital in Paris and worked with him to test the effects of radium on resistant lupus as compared with conventional treatments. There was little difference. Pierre thought radium might cure blindness. He and Becquerel charted the work of Dr. Jauel, an eminent eye specialist, as he applied the radium treatments, which failed to help. By 1904 Pierre and Marie were involved with several technicians and doctors in testing radium on numerous illnesses from cancer to tuberculosis. The results were poor.

The first recorded successful use of radium treatments to cure cancer had taken place the previous year not in France but in St. Petersburg, where two patients suffering from basal cell carcinoma of the face were successfully treated. In that same year, Alexander Graham Bell proposed using radium to treat diseased internal tissues. At his suggestion a French doctor, Henri-Alexandre Danlos, developed a glass capsule containing radium salts, which could be inserted into the cervix or uterus. By 1904, Parisian doctors working with the Curies' radium samples used these glass capsules as well as a new device that consisted of a hollow tube with a small cup at the end that contained radium-impregnated varnish. This treatment was successful and began demonstrating that radium could be more effective than X-rays in treating deep-seated internal tumors. Danlos wrote, "Radium, though costly, is a very convenient substance at a point where the application of X-ray is impossible or exceedingly difficult."

Although the scientific world was slowly becoming aware of the efficacy of radium treatment, its use was extremely limited. It seems ironic that the Curies themselves retarded the medical use of radium. From 1904 to 1906, while the Curies were not affected by the Austrian embargo, other scientists were prohibited from buying radium at a reasonable price and the Curies stopped lending samples. A bitter Dr. H. Strebel of Munich reported that he had inserted radium enclosed in a capsule of paraffin wax into a woman's uterus (a method superior to the glass capsule used in France, since it could not break inside the body), but that his results were extremely poor due to "the low radiation capacity of German radium available to me."

During the embargo, radium had become so costly that modestly priced X-ray therapy grew rapidly as doctors and technicians improved their methods of application. In the first textbook on X-ray and radium treatments, published in 1904 in Berlin, only 17 of the 306 pages addressed to technicians concern radium therapy. In 1906, when the Austrian embargo was lifted and radium became less expensive, treatment research resumed. That year a New York surgeon, Dr. Robert Abbe, successfully sealed radium's emanation, radon gas, in a celluloid capsule (to avoid breakage) and placed it directly into diseased tissue where X-ray could not reach, thus obtaining the first successful American cures.

Nevertheless, most gynecologists and internists doubted the effectiveness of radium treatments. When Dr. C. J. Gauss, an expert radium therapist, lectured to distinguished doctors and medical technicians at the Berlin Society and spoke of the "positive results" he had secured through radium treatments, he was booed, but such treatments for breast, lung, and rectal cancers became more and more successful. To treat breast cancer, approximately fifty needles containing up to

100 milligrams of radium bromide were inserted directly into breast tissue. Radon "seeds" or small pellets also were used but often severe reactions occurred. In 1916, Dr. Gioacchino Failla at New York's Memorial Hospital added a gold filter, thus making the treatment more viable. In the hospital's basement, laboratory workers prepared the "seeds." Instead of disposing of the gold filter after each use as instructed, one laboratory worker took the gold and forged it into a wedding ring for his fiancée. Eventually she had to have her ring finger amputated.

As radium treatments became more common, research forged ahead. Various treatment systems were developed in Munich, Stockholm, and Manchester. At the Curie Institute in Paris, cork applicators of various sizes and hard rubber ellipses called "ovoids" were used. At Memorial Hospital in New York, there was the "vaginal radium bomb" consisting of a lead sphere supported by a rod for insertion.

Later, at the Curie Institute, Marie and Irène were to work on standardizing radium treatments. By 1922, in France, the same dose of radiation was still recommended for every patient, but in England and Belgium, therapists were working on individualized measurements to pinpoint specific areas. By 1934, Herbert Parker and James R. Paterson devised the Manchester system, which provided an exact dose of radiation. By reusing radium and milking it for its radon gas, the cost of treatment became reasonable. Starting in 1935, radium became the preferred material for treatment. Although it provided the deepest penetration, the radiation sessions were often extremely long. Dr. R. F. Mould described this process as used at London's Westminster Hospital: The output of the typical radium bomb was so low that the patients, protected

by a lead apron, were often given books, newspapers, and cig-
arettes to occupy them. In the middle of the treatment, they
were allowed to leave the treatment couch and stroll around
the room. On the wall a puzzling sign sent a mixed message:
"This room is perfectly safe for patients. It is inadvisable for
those working constantly with radium to spend more time in
this room than is necessary."

As sources became more abundant, radium therapy continued to be the substance of choice until the middle 1950s,
when cobalt replaced it. Cobalt-60, a radioactive isotope, is a
strong source of gamma rays and therefore is effective for use
in treating cancers as well as in medical tracers. Treatment
planning, measurements, and the design of apparatus
improved steadily. In the 1960s, the Manchester system was
superseded by the French Pierquin system and later the
advent of computerized treatment-planning software. Today
a complicated array of substances and treatment modalities
has made the exclusive use of radium obsolete. The term *Curietherapy* has disappeared, but the name Madame Curie and
radium remain forever inseparable.

CHAPTER 20

To Pass the Torch

The one satisfaction in Marie's life seemed to have been her close working relationship with Irène, and she determined that after her death the control of the Curie Institute would go to her gifted daughter. These two shared a consuming interest in science. Eve was shut out and as an adolescent she began to rebel. After a summer vacation when Eve was sixteen, Marie wrote Irène, "I hope . . . our Evette will love us more in Paris than she did in l'Arcouëst." When Eve practiced the piano hoping for a concert career, her mother referred to it as "noise." "We will have to reconcile the scientific work which we two represent with the musical art represented by Evette, which is much easier to do in good weather than in rain," she wrote Irène. While her elder sister used no makeup and often wore a white laboratory coat over her inexpensive dark dress and nurse's shoes, Eve, dressed in fashionable clothes, experimented with kohl eyeliner, bright red lipstick, and a variety of other cosmetics which appalled her mother. Charm and grace counted for little in her family.

In 1925, a nervous young man of twenty-five, Frédéric Joliot, still wearing his uniform from the antigas corps, presented himself to Madame Curie. As a boy he had kept a photograph of the Curies taped to his bedroom wall. Madame Curie looked up from the papers she was reading and after a brief conversation asked, "Can you begin work tomorrow?" "I have three weeks of service to complete," he answered. "I will write your colonel," she said and with that went back to her reading. The interview was not as casual as it seemed, for Paul Langevin had highly recommended Joliot. He began work the next day.

Frédéric's father, Henri Joliot, had fought for the Paris Commune and was exiled from France until amnesty was declared. A successful businessman, Henri played the French horn and was an avid fisherman. After his father's death, Frédéric enrolled at the School of Industrial Physics and Chemistry of the City of Paris (EPCI), where Pierre Curie had taught before turning over his post to Paul Langevin. Joliot studied under Langevin and adopted many of his personal gestures as well as his left-wing politics. "I owe my basic culture and knowledge to Langevin," he said.

Frédéric was an extrovert, and quickly established friendships. In the silent, austere atmosphere of the Curie laboratory he found himself very much alone. He steered clear of Irène, "la Patronne's" favored assistant who never said "Good morning" and, as Frédéric observed, "did not always create a feeling of sympathy around her at the lab." But gradually he was drawn to her intelligence and found in her "a living replica" of Pierre Curie. After a year of long walks and intimate talks, one evening Irène came home and announced, "Mé, I'm engaged." Marie was devastated. Her partner, friend, and protector was about to leave her for what she thought was a marriage of convenience. Irène too worried that her mother thought that Frédéric was taking advantage of the Curie name and accomplishments to further his career. Few could picture this handsome, chain-smoking ladies' man with the taciturn Irène, who at 28 was three years his elder and had declared that she would never marry. Her mother tried to discourage the match. She insisted on an agreement that voided the French law that husbands controlled their wives' property. Marie consulted a lawyer to make sure Irène alone would inherit control of the Curie Institute's radium and other radioactive substances.

Irène proved to be as stubborn as her mother. On October 9, 1926, Irène and Frédéric married at City Hall in the fourth *arrondissement*. Eve prepared a celebratory lunch and then the newlyweds returned to the laboratory. For the first few months of their marriage they lived with Marie, but as soon as they were able they moved into their own apartment on rue Froidevaux in the building jointly owned by Marie, the Perrins, and the Borels. Three nights a week they dutifully had dinner with Marie, but for two years she introduced Frédéric as "the man who married Irène." In an almost eerie re-creation of her behavior with Pierre, Marie insisted Frédéric take his second baccalaureate, then his bachelor's degree and doctorate. He obeyed. He progressed from being a far inferior chemist and physicist than his wife to excelling in both fields. His charm and scientific progress won Marie over. Three years later she told Langevin, "That boy's a fireball."

Irène signed her scientific papers "Irène Curie," and he signed his "F. Joliot," but soon adjusted both their surnames to "Joliot-Curie." In 1927, a daughter, Hélène, was born.

Shortly after, Irène's doctor informed her that she had contracted tuberculosis and warned her not to have another child
and to cut back on work. The next week Irène was back in the
laboratory and five years later gave birth to a son, Pierre.

Thanks to Marie's efforts the Joliot-Curies had almost two
grams of radium at their disposal at the Curie Institute from
which they isolated 200 millicuries of polonium, the most
powerful source of alpha rays. This was more polonium than
at any other laboratory in the world and said to be ten times
stronger than German polonium. Marie had passed the torch
to the next generation and provided them with the material
they needed to further explore radioactivity and the nucleus
of the atom. Once again France and the Joliot-Curies took the
lead in science, with Lise Meitner in Berlin, Ernest Rutherford in England, and Niels Bohr in Copenhagen pursuing the
same subjects. "We had to speed up the pace of our experiments for it is annoying to be overtaken by other laboratories
which immediately take up one's experiments," Frédéric wrote.

In 1930, two German scientists, Walther Bothe and Herbert Becker, observed that while bombarding elements such
as beryllium with alpha rays, a powerful radiation was emitted, one that could pass through ten centimeters of lead. Irène
and Frédéric, like other scientists of the day, assumed that
the atom contained only protons and electrons (negatively
charged particles). They repeated Bothe and Becker's experiments and were convinced that highly energetic, deeply penetrating radiation emanated from a new type of gamma ray
that moved at the speed of light. They published their findings in January of 1932. On reading their paper Rutherford
exclaimed, "I can't believe they're gamma rays." Rutherford
suggested that one of his young researchers, James Chadwick,

pursue the question. He borrowed some polonium from Lise Meitner at the Kaiser Wilhelm Institute and from an American hospital, repeated the experiments and discovered neutrally charged particles inside the nucleus, which he named "neutrons." Chadwick won the Nobel Prize for discovering the neutron. Frédéric and Irène, who had produced the same results but interpreted their own data incorrectly, lost out. It would not be the only such failure.

Frédéric began to study neutrons in a Wilson cloud chamber, which recorded the trajectory of particles in tiny droplets of condensation. He mounted two cameras (so he would not be interrupted by having to reload his film during an experiment) above the chamber to record these movements and stood for eight hours a day photographing in order to establish a pattern. The droplet pattern seemed odd to the Joliot-Curies; it looked as if some sort of positive particle was the same size as a negatively charged electron. Frédéric and Irène concluded that somehow a negatively charged electron had entered the cloud chamber and headed straight for the neutrons. In the summer of 1932, a Californian physicist, Carl David Anderson, discovered a new particle. He had exactly replicated the Joliot-Curies' experiments and reached the correct conclusion. What the cloud chamber had shown was the existence of an electron with a positive charge, which Anderson named a "positron."

Frédéric had been alone in l'Arcouëst when this discovery was made because Irène had stayed in Paris to rest after a hard winter of working with radioactive substances. She attributed her increasing loss of energy solely to her tubercular condition. When Frédéric returned to Paris in September, he restudied his own cloud chamber photographs and clearly saw the positron. The Joliot-Curies had missed a second chance for a Nobel Prize. Using their polonium's strong alpha rays, Frédéric and Irène began to bombard various elements looking for more positrons. Their findings surprised them, in that lighter elements sometimes ejected a neutron followed by Anderson's newfound positron, instead of the expected proton.

In October 1933, the world's most eminent physicists met for the seventh Solvay Conference in Brussels. The great scientists included Niels Bohr, Enrico Fermi, Frédéric Joliot-Curie, Wolfgang Pauli, Ernest Rutherford, Werner Heisenberg, and Paul Langevin. There were three women scientists: Marie Curie, Irène Curie, and Lise Meitner. This was the first conference dedicated to a discussion of nuclear physics. The Joliot-Curies presented a report, "Penetrating Radiation from Atoms Bombarded with Alpha Rays," in which they told of their findings of the unexpected ejection of a neutron and a positron. A debate ensued in which Meitner attacked the young team. Meitner, "the iron lady" of physics, was also a masterly chemist and mathematician. She was rarely wrong. She announced that she had performed several similar experiments at the Kaiser Wilhelm Institute and had produced no such results. The debate turned into a denunciation of the Joliot-Curies findings. Irène and Frédéric left the meeting "feeling rather depressed." Marie, who looked emaciated, her face a mask of deep wrinkles, her hair white, had not taken part in the debate, and she turned away when Rutherford tried to engage her in conversation. She glared at Lise Meitner, but said nothing.

By 1934 Meitner, who had hidden under the seats at Berlin's Chemistry Institute in order to hear lectures forbidden to women, had evolved into a powerful presence. Her division of the Kaiser Wilhelm Institute was well funded by the German government and by the I. G. Farben Company, soon to become infamous for its use of slave labor and the construction of the Nazi gas chambers. Meitner was not only a partner of Hahn's, but the dominant one. She worked out the mathematical calculations for their experiments and longed to be the first scientist to reveal all the components and activity of the atom. But Meitner was racing against an implacable foe, one far greater than any scientific challenger.

At the end of January 1933, Adolf Hitler became chancellor of Germany, and on March 23 the Enabling Act granted the Reichstag's powers to Hitler's cabinet. German Jews were deprived of their civil rights, and a national boycott of all Jewish businesses was imposed. Riots began in which mobs attacked Jews. A week later another act was passed wherein "non-Aryans" (the definition included any person with at least one Jewish grandparent) were to be purged from all government agencies and universities. Meitner, an Austrian whose family had converted from Judaism, regarded herself a Protestant and felt protected from this edict. Jewish scientists were being fired or were leaving of their own accord, but Meitner felt herself "too valuable to replace." Niels Bohr sensed what was to come and obtained a year's Rockefeller grant for Meitner at his institute in Denmark. Otto Hahn and Max Planck talked her out of it, but, in September 1933, she was officially informed that she no longer had the right to teach at the University of Berlin, publish her articles, lecture, or attend scientific conferences in Germany. As she arrived at the Solvay Conference the following month, she felt that the institute she "built from the first stone" was slipping away from her, while Madame Curie and the upstart Joliot-Curies, in no danger, had the support that was prohibited to her.

Before Irène and Frédéric left the Solvay Conference, Bohr and Pauli took them aside and encouraged them to ignore Meitner's attack and to repeat their experiments. Perhaps Meitner had missed something. Marie's credo, one that had put her under great pressure, was "You must get so that you *never* make a mistake. The secret is in not going too fast." Lise, on the other hand, assailed Irène, saying that as a scientist she was using "her mother's old fashioned methods" taking a "methodical . . . long and slow path" that had become outdated. In Paris, the Joliot-Curies repeated their experiments with the same results. They thought that perhaps the neutrons and positrons they observed might be caused by the energy of the alpha particles from their polonium source attacking their aluminum-foil target. Carrying the experiment further, they moved the alpha bombardment both nearer and farther away from the target. Frédéric observed that at a short distance the alpha bombardment did produce neutrons, proving Meitner wrong. When they increased the distance between the alpha rays and the target, as might be expected, the neutron emissions stopped.

Yet here, as in Marie Curie's discovery of natural radioactivity, the unexpected occurred. The positron emissions did not immediately cease but slowly decreased in a manner identical to the radiation from naturally radioactive substances. Frédéric picked up a Geiger counter and repeated the experiment. The counter clicked away. He then removed the alpha source but the counter did not become silent; it kept on clicking until it lost its intensity a full three minutes later. Aluminum was a stable element and yet the aluminum foil was registering radioactivity. Was the Geiger counter defective? Before the Joliot-Curies left the laboratory they asked that the counter be checked. It was in working order. The Joliot-Curies had discovered how to create artificial radioactivity.

There was no way to check their discovery chemically because of the infinitesimal amount of aluminum involved. However, if they had been successful some of the aluminum would have decayed to radioactive phosphorus. This would have to be proved within its three-minute half-life. To effect this, the Joliot-Curies at top speed irradiated the aluminum foil, dropped it in a solution of hydrochloric acid, and quickly stoppered the tube. The acid dissolved the foil but left the phosphorus. They drew this gas into another tube. Using a Geiger counter, the aluminum itself was silent, but the phosphorus gas activated the counter. They had produced chemical proof of their discovery. *Comptes rendus* in the January 1934 issue called artificial radioactivity "one of the most important discoveries of the century." The day of the discovery, Irène and Frédéric set up the experiment once again. Late in the afternoon, Marie Curie and Paul Langevin entered the laboratory. Two years previously, Irène had taken over as director of the Radium Institute; her mother was mortally ill but still fighting to continue some laboratory work and update her two-volume *Treatise on Radioactivity*. Frédéric began the experiment, explaining as he went along:

> I will never forget the expression of intense joy which came over her [Marie] when Irène and I showed her the first artificially radioactive element in a little glass tube. I can still see her taking in her fingers (which were burnt with radium) this little tube containing the radioactive compound—in

which the activity was still very weak. To verify what we had told her, she held it near a Geiger-Müller counter and she could hear the rate meter giving off a great many clicks. This was doubtless the last great satisfaction of her life.

As Marie's body deteriorated, various diagnoses were given: old tubercular lesions, gallstones, liver and kidney damage. She coped with tinnitus, a continual humming in her ears; she had endured two painful operations for cataracts but was still nearly blind. She refused to take the blood tests she prescribed for her workers, but her pallor and lack of energy suggested anemia.

Marie visited her laboratory one more time in May. She was frustrated when an experiment did not go as planned and left the laboratory at 3:30 P.M., complaining of a headache, fever, and chills. As she stepped outside she noticed that several roses in her garden were fading and instructed the gardener to take care of them. Marie was dying. Eve, the child she never understood, became her nurse and devoted companion. She accompanied her mother to Sancellemoz, a sanatorium in Haute-Savoie Mont-Blanc in the French Alps. Marie's condition became steadily worse and her temperature climbed, but on the morning of July 3, 1934, the thermometer registered normal. Eve felt hopeful of a rally. Soon after, Marie lapsed into a coma. She died at dawn. Her last words were, "I want to be left in peace." The doctor said that it was something of a miracle that she had lasted to 67, and gave the cause of death as "aplastic pernicious anemia. . . . The bone marrow could not react probably because it had been injured by a long accumulation of radiation." Marie was buried next to Pierre in Sceaux and remained in this grave for sixty-one years until both bodies were exhumed and transferred to the Panthéon.

Even on her deathbed, Marie had insisted that a dose of fresh air might be all that was needed to help her recover. With the persistence that had allowed her to perform seemingly impossible tasks, Marie Curie never acknowledged that her beloved radium might have betrayed her. It was the same denial with which she proclaimed, though once a star pupil in Warsaw who had performed flawlessly in Russian, that she could neither speak, read, write, or understand that hateful language.

A frequently asked question is, How could her denial have been so strong? How could the Curies expose themselves, their associates, and even their precious daughter Irène and her husband to the devastating effects of radiation? The answer, I believe, was love. It prevented Marie and Pierre from seeing radium with the same cold, scientific eye they brought to their other work. Even as they warned of the dangers of radium exposure, at their bedside the Curies had kept a vial of radium salts to observe its beautiful glow before falling asleep. Marie referred to radium as "my child."

Along with love no doubt was Marie's intense belief that great scientific discoveries demanded sacrifice. From childhood she had been inculcated with the theory that deprivation and disregard for personal welfare in the service of a great cause were noble characteristics. Pierre had shared her view, and so they overlooked, or ignored, or blocked from their consciousness the danger they faced from their hard-won discovery. This attitude is difficult to accept even if one takes into account that the deleterious effects of radium could neither be seen nor felt at the time of exposure. As early as 1903 in his Nobel lecture, Pierre obliquely referred to Becquerel's accidental burn after placing the vial of radioactive barium salts in his vest pocket:

> In the biological sciences the rays of radium and its emanation [radon gas] produce interesting effects which are being studied at present. . . . In certain cases their action may become dangerous. If one leaves a wooden or cardboard box containing a small glass ampulla with several centigrams of radium salts in one's pocket for a few hours, one will feel absolutely nothing . . . [but] a prolonged action could lead to paralysis and death. Radium must be transported in a thick box of lead.

This from a man whose own fingers and those of his wife had become hard as cement with recurrent fissures that split open like red crags in clay. Both Marie and Pierre had experienced a numbness in their fingers. Marie developed a habit of constantly rubbing her deadened finger tips with her thumb as if trying to restore some shred of sensation. In 1904 Thomas Edison's assistant had died of radiation poisoning while trying to develop an X-ray lightbulb. By then the Curies knew that X-rays were a great deal less harmful than radium rays.

At the Kaiser Wilhelm Institute, Lise Meitner insulated her laboratory with lead and warned of the dangers of exposure to radioactive substances. In a primitive form of protection, Meitner mandated frequent hand washing and placed toilet paper next to every door handle insisting it be used when opening and closing doors. The lecture rooms had dark and light chairs—the light chairs for those who worked with weak radioactive substances, the dark for those who studied stronger radioactive materials. Meitner installed fans and hoods over work tables to carry away the fumes and housed radioactive materials in lead boxes. She required her staff to use forceps when handling radioactive materials. In a display of "do as I say, not as I do," Marie also installed many of these protections, but both she and Irène largely ignored them. They used their naked hands in experiments and shockingly often transferred radium and polonium from one vessel to another by sucking up these substances with a pipette. Over the years, even as they grew sicker they continued to work unprotected.

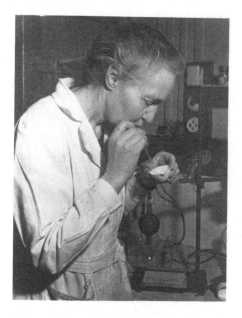

In her laboratory, in 1954, Irène Joliot-Curie sucks on a pipette to transfer dangerous substances.

During World War I, when both mother and daughter had worked at various field hospitals, they were exposed to massive doses of X-rays and radon gas. In 1921, Marie Curie wrote in *Radiology and the War* that radiodermatitis could lead to death. She took little personal notice of her own findings. The year 1925 marked the turning point in denial. Nonmedical and industrial uses were obviously posing danger. At wooden tables in the U.S. Radium Corporation factory in New Jersey a row of young women sat painting luminous numbers on watch dials, diligently licking their brushes to bring them to a fine point. The paint contained only one part radium per six-hundred thousand parts of inert substances, yet within three years, fifteen young women perished as radium poisoning destroyed their jaws and bone marrow.

That same year in Paris, two engineers who were former pupils of Marie died after preparing industrial solutions of thorium X. Another had his fingers amputated, then his hand, then his arm. Subsequently, he went blind. Soddy maintained that radium had rendered him sterile. Soon deaths among radiologists, industrial *préparateurs*, radium researchers, and the public at large became so common that the French Academy of Sciences classified the process of radium's manufacture as "a highly dangerous operation." Madame Curie warned that industrialists and engineers must use protective measures and take blood tests, but she noted that there had been absolutely no serious effects or "accidents" at her own institute.

In her autobiography Marie Curie admits that radiation may have damaged her health, but only in a minor way: "Since the handling of radium is far from being free from danger (several times I have felt a discomfort which I consider results from that cause), measures were taken to try to prevent harmful effects in the persons preparing emanations." *Discomfort* indeed. There was no doubt that radium had destroyed her, slamming its raging power into her bones and organs. A century later, this contamination still clings to the preserved clothes she wore. But when she wrote to Bronya of the intensity of the humming in her ears which caused her acute distress and of her deteriorating eyesight, she speculated, "Perhaps radium has something to do with these troubles, but it cannot be affirmed with certainty."

When Irène Joliot-Curie died at fifty-nine in 1956, her death was duly noted as leukemia induced by exposure to radioactive substances. It was commonly thought, because the names *radium* and *Curie* had become synonymous, that Irène died solely from radium exposure. The main cause, however, was her youthful World War I exposure to X-rays and radon exacerbated by a capsule of polonium-210 that exploded on her laboratory table fifteen years before her death. This deadly substance is readily absorbed into tissue and is too dangerous to handle even in minuscule amounts. Irène's husband, Frédéric Joliot-Curie, who was so weak himself that he could only briefly visit his wife's bedside, died two years later from the effects of radium and polonium. With morbid humor, Frédéric called death from radioactive exposure "our occupational disease."

CHAPTER 21

Marie's Legacy

Patriotism, sacrifice, the quest for peace, the equality of women, excellence, all this constituted Marie Curie's legacy. By 1935, thirty-two years had passed since Marie had won her first Nobel Prize and in all that time it had gone to no other woman scientist. She had not lived to see her daughter Irène be the next woman scientist to receive the Nobel Prize along with her husband for their discovery of artificial radioactivity.

Some things had not changed. The press coverage almost universally attributed the prize to Frédéric's talent while Irène was relegated to an assistant's role. But the Joliot-Curies did share the Nobel speech and Frédéric began his part with a tribute to Marie and Pierre's work that had eventually led to their own discovery. Near the end of this speech he too expressed a fear that "scientists building up or shattering elements at will, will be able to bring about transmutations of an explosive type."

After returning to Paris the Joliot-Curies began to investi-

Irène and Frédéric Joliot-Curie receive the 1935
Nobel Prize from King Gustav V of Sweden.

gate the artificial creation of new radioactive elements. They
were quickly followed by Niels Bohr in Denmark, Enrico
Fermi in Italy, and Lise Meitner, who had remained in Berlin.
Although Meitner had not worked with Otto Hahn for a
decade, he joined her in this project. The third member of the
team was Fritz Strassmann, who had refused to join the
National Socialist Party. As a result of his opposition to Hitler,
Strassmann found himself unemployable, penniless, and
starving. At Meitner's suggestion Hahn hired him for fifty
marks a month, all of which went for food.

Hahn observed that scientists were now creating artificial

elements "like the old days when new elements fell like apples from a tree." These new elements were referred to by most scientists of the day as *transuranics*. In Italy, Fermi, who had written the definitive study on beta decay, also began to bombard heavy elements but in the course of his experiments found that he achieved better results using neutrons than alpha particles. Fermi then observed that counter to one's instinct if the neutrons were slowed down they became even more effective. In this manner, within the next three years scientists in France, Germany, Italy, and Denmark discovered what they thought to be more than four hundred new transuranic elements. Rutherford, because he had trained Hahn, never questioned the true nature of these discoveries. With time he might have done so, but in October of 1937 he died unexpectedly after a minor hernia operation.

Enrico Fermi was particularly interested in uranium, the heaviest of the natural elements. He bombarded uranium with slow neutrons and succeeded in producing what he was convinced were new artificial elements, heavier than any known element. He named them "ausenium" and "hesperium." In December of 1938, Fermi stood before King Gustav V of Sweden and in the presence of his proud family received the Nobel Prize for the discovery of two new elements—elements that in fact did not exist.

After his award Fermi firmly shook King Gustav's hand. The king was taken aback because Fermi had committed a serious breach of conduct. He had not given the required Fascist salute. Fermi, a Catholic, was married to Laura Capon, the daughter of an Italian naval officer. They had two children: Nella, age seven, and Giulio, who was two. In Italy the summer of 1938 marked the "Manifesto della Razza," which declared that Italians were Aryans, except for Jews. Laura was Jewish. Fermi determined that his family must escape from Italy, but with a "non-Aryan" wife and children it was almost impossible to do so. Also, even if they could find a way to emigrate, they could take almost no money with them. Niels Bohr came to the rescue. In the early fall of 1938, Bohr was chairman of an annual gathering of scientists in Copenhagen. On the first morning of the conference, Bohr spoke out against Hitler and his policies. The German scientists stood, gave the Nazi salute, and walked out. Later that day, Bohr took Fermi aside and whispered to him the secret news that Fermi was among those nominated for the Nobel Prize. In an indirect way, Bohr indicated that Fermi was sure to win and that the prize money, then worth about a hundred forty thousand dollars, could be deposited in a United States bank. If by December 10th Fermi could obtain six-month tourist visas for his family, Bohr thought he need not return to Italy. When Fermi received his official award letter, he wrote the Nobel Committee that he would like to have his wife and children attend the ceremony. The escape plan went smoothly and a few days later the Fermi family boarded the *Franconia* headed for New York.

It was not so easy for Lise Meitner. With the *Anschluss*, when Germany annexed Austria in a bloodless coup, Meitner was no longer Austrian, but a Jew in Nazi Germany. Hahn, fearing the loss of his own job, instructed her not to enter her own institute. Meitner was now isolated. She tried to leave the country but was turned down with the comment, "It is considered undesirable for well-known Jews to travel abroad." Then came news of her imminent arrest. Meitner packed a few summer dresses and with only ten marks in her pocket prepared to escape. At the last minute, Hahn appeared and gave her a diamond ring that had belonged to his mother to use in an emergency. She boarded a train to the Netherlands using her invalid Austrian passport. She was not detained. For a time she worked with Bohr in Denmark; from there she went to the Manne Siegbahn Institute in Stockholm, but she was given no equipment and therefore could not conduct experiments. "I feel so completely lost and helpless," she wrote. "I am gradually losing all my courage. . . . One dare not look back, one cannot look forward." Her only joy was to correspond with Strassmann and Hahn as they continued the work she had started.

Irène Joliot-Curie had painstakingly replicated Fermi's and the German scientists' experiments after they had supposedly found a variety of new elements heavier than uranium. Irène suspected that these so-called transuranic elements were not what they seemed, although what they were, she did not know. She then wrote a paper stating that these scientists were probably mistaken. As a result, Hahn and Strassmann performed experiments hoping to prove her wrong. In December, Hahn wrote to Meitner of an odd occurrence: they had bombarded a uranium target with slow neutrons, but instead of losing a few particles as expected, the bombardment produced elements resembling barium, which weighed about half as much as uranium. "Perhaps you can propose some fantastic explanation," he wrote.

Meitner was spending Christmas with her nephew, Otto Robert Frisch, who had escaped from Germany and was working with Bohr. His father was in the Dachau concentration camp. Frisch and Meitner went out on a snowy morning, he on skis, she walking beside him. They sat on a tree trunk while Meitner started to calculate on scraps of

paper. Her calculations showed that a heavy unstable uranium atom could be deformed to resemble an attenuated
drop of water that formed a figure eight (or a dumbbell)
which then could split into two so-called liquid drops, sev-
ered by a mutual electrical repulsion of great magnitude.
Also, she knew the exact mass of uranium and of barium,
and realized that some of this mass disappeared in the splitting process. Meitner calculated that one gram of uranium
contained the unbelievably huge number of 25 followed by
twenty zeros. By using Einstein's formula of relativity in her
calculations, she found that the energy discharged when the
mass disappeared and that which was set free when the
atom split were roughly the same. The immense energy
released in the splitting process was to set atomic science
on a new path.

In fact, what illustrious scientists had been doing for the
four years since Irène and Frédéric had discovered what they
called "artificial radioactivity" was not finding new transuranics as they thought, but unknowingly splitting the atom. They
had created isotopes of existing elements. When Irène Joliot-
Curie read a paper by Hahn and Strassmann explaining that
the atom had broken in two (Meitner and Frisch, who named
this process "fission," were Jews and therefore banned from
putting their names on any publications), this usually unflap-
pable woman exploded, "Oh, what dumb assholes we've
been!" It was clear that Meitner had succeeded while others
failed in solving the mystery of nuclear fission. In 1944, Otto
Hahn became the sole winner of the Nobel Prize in Chemistry for this discovery. There was no mention of Lise Meitner. It was not until 1992 that German physicists fused
isotopes of iron and bismuth to create the heaviest artificial

element then known. They named it "meitnerium" in honor of this forgotten scientist.

Hélène Langevin-Joliot says,

> In discovering artificial radioactivity my parents did exactly what my grandmother had done but in reverse. In each case, they created a situation in which the science had to be rethought. Marie upset the apple cart by finding radioactivity and then Irène did the same thing by finding atomic fission, only she didn't know what she had done. Theoretically it could not exist, but chemically it did.

Niels Bohr learned of nuclear fission just before boarding a ship for the United States. "What idiots we have all been," he exclaimed. One week after Bohr arrived in America, on J. Robert Oppenheimer's blackboard was a crude sketch of an atomic bomb.

The Joliot-Curies, imbued with Marie's love of peace, envisioned only the good that might come from their discovery. Nuclear power might supply France with much-needed energy instead of importing coal, oil, and such products. In 1939, Frédéric became a captain in the French army in charge of a group of scientific researchers whose work involved finding the exact amount of energy released in a chain reaction. Afraid that his findings might be used by the Germans to make a bomb, he deposited the results of these experiments in a secret bank vault rather than publishing them. In bombarding uranium atoms to create a chain reaction, an ideal substance was heavy water, a combination of oxygen and deuterium that slowed the neutrons so that they reacted more easily with the uranium. The largest store of heavy water was manufactured at the Norsk Hydro Company in Norway. The Germans had made overtures to Norway, a neutral country, to secure this substance for their weapons program. Frédéric, sensing the danger, arranged to smuggle Norway's supply of 185 kilograms of heavy water to Paris by airplane.

When the Germans advanced on Paris, spies informed them of this secret supply. By the time German officers arrived at the Curie Institute, Irène and Frédéric were headed for the south of France. The officer in charge of the operation met them, his car loaded with the cans of heavy water. Frédéric was given a metal tube of cadmium. If the mission failed, a small amount of the cadmium would instantly render the heavy water useless. In constant danger of betrayal, Frédéric and two companions traveled by day and locked the precious cargo in the cells of local French police stations at night. The heavy water finally was loaded onto a ship sailing for Southampton, England. It arrived safely. Then Frédéric "disappeared." Under the pseudonym Jean-Pierre Caumon, he joined the French Resistance.

The high-spirited Eve Curie also joined the Resistance. Her knowledge of German helped the Allies. She became a war correspondent and codirector of the newspaper *Paris-Presse*. Like her mother, patriotism meant all to her, danger nothing. After the war she married Henri Labouisse, who became the director of UNICEF (United Nations International Children's Emergency Fund). Much of the year was spent traveling around the world seeking to better the plight of children. In 1965, Henri Labouisse too was awarded the Nobel Prize on behalf of UNICEF's efforts for worldwide peace.

Irène, separated by war from her husband, refused to leave France until her daughter Hélène, then sixteen, had completed her baccalaureate. She took her examination in secret in a small French village. Irène's son, Pierre, who was then eleven, would later become a professor at the College of France in charge of the Chair in Cellular Bioenergy, and a professor at the École Normale Supérieure in Paris. He is a recipient of the Legion of Honor.

On June 6, 1944, Irène and her two children strapped knapsacks on their backs and began a perilous hike over the French Alps into neutral Switzerland. Luckily, unbeknownst to them, German guards were paying little attention to the border that day. It was D-day, the day of the Normandy invasion.

During the war Paul Langevin was arrested by the Gestapo and charged with Allied collaboration. His son-in-law was shot, his daughter was deported. In July of 1944, Langevin escaped and fled to Switzerland. The Perrins managed to leave occupied France and came to the United States. Jean Perrin died the next year. The old circle was no more.

At the war's end, Frédéric, who emerged a hero of the Resistance, became one of the most influential men in France. He had joined the Communist Party, which at that time was a powerful anti-Fascist group, and was put in charge of the French Atomic Energy Commission. As for Irène, Missy Meloney sent her streptomycin, which cured her tuberculosis. In 1936, under the new Léon Blum government, Irène had been appointed Undersecretary of State for Scientific Research, a position she held briefly, saying that she had only accepted it to further the cause of women in France. Irène inspired women with her lectures, both in France and abroad, advocating the rights of working mothers, the protection of children, and the goal of world peace. In 1945, she was instrumental in at last gaining the vote for the women of France.

Irène and Frédéric, like Marie and Pierre, applied for membership in the Academy of Sciences. He was accepted. She was rejected. Unlike her mother, however, she applied twice more, transforming each rejection into a continuing battle for women's rights. Like her mother, Irène resented the time and effort that went to seeking funds to further science. "We don't realize that science is part of our most precious patrimony . . . which advances human life and decreases suffering. May the public make the future easier. . . . This cry of Marie Curie still retains its value."

As the cold war began, the attitude toward French communists shifted. Irène told her daughter Hélène that as a Curie she could never join any political party, but she strongly believed in peace and that "nuclear energy has only one objective, the improvement of the economy of our daily lives." Marie had instilled these beliefs in her daughter. In 1945, when the atom bombs had fallen on Hiroshima and Nagasaki, Irène said that she was glad her mother had not lived to see that day.

Frédéric's Communist Party affiliations, however, brought him increasingly in conflict with the government. As head of the French Atomic Energy Commission, in 1950 he was asked his position on nuclear fission. He replied, "I believe that in order to defend the peace by peaceful and effective means, we must translate our will into actions. . . . If tomorrow we were asked . . . to work on the atomic bomb, we must reply—No!" Frédéric Joliot-Curie was summarily dismissed.

Today a third generation of Curie women carry the torch. Hélène Langevin-Joliot graduated from the EPCI at seventeen and immediately went to work as a researcher for her father, who was developing France's first atomic reactor. "In those days school was a lot shorter and training longer," she said. "The idea of bringing energy and economic improvement to France had long been my father's goal." France was then importing almost all of its fuel oil and over 35 percent of its coal.

On December 15, 1948, two years before Frédéric Joliot-Curie's dismissal, France's first atomic reactor, ZOE (Z for zero, as the power was small; O for oxide of uranium; E for *eau lourde,* or heavy water) had been activated. Hélène stood nervously among the crowd while the needles on the gauges quivered then shot up as the pile reached critical mass. What Frédéric Joliot-Curie had started was to lead to a France where 80 percent of its electric energy is generated within this country and the surplus is exported.

In the small, tightly knit scientific community in France it was not considered unusual that Hélène married Michael Langevin, the grandson of Paul Langevin. In 1950, their daughter, Francine, was born and the next year a son, Yves. Once again a Curie woman set out to balance motherhood and a career. In 1954, Hélène helped develop a scintillation spectrometer. Two years later she defended her thesis, and by the end of the 1950s she had established herself as a leading scientist in France with her study of the polarization of electrons emitted through decay.

The age of the individual titans of radioactive science was drawing to a close. Teams of scientists began working together. Hélène has spent her career in particle physics and the study of heavy nuclei. Now computers link far-flung laboratories, and Hélène Langevin-Joliot's papers most often reflect the collective work of as many as twenty scientists working worldwide.

The year before her father's death in 1958, Hélène joined the faculty of the Institute of Nuclear Physics at the University of Paris at Orsay, where she became director of research in charge of the 580-person laboratory and the synchrocyclotron, which was modernized under her direction. Like her parents and grandparents, Hélène is a pacifist; and like all three generations of the Curie women she lectures on the importance of genderless education and women's rights. She would like to see more women in science. "In a photograph of the 1933 Solvay Conference, there were three women scientists— Marie Curie, Irène Curie-Joliot, and Lise Meitner—and thirty-five men," she notes. "If they were to take a photograph today the percentage wouldn't be much different."

And now the author must inject herself into this book. Shortly before I left France, Hélène Langevin-Joliot, the third generation of remarkable Curie scientists, invited me to visit her home in Sceaux, the one she had inherited from her parents. Here the past comes vividly alive. The house seems caught in a time warp. A display cabinet full of memorabilia from the Joliot-Curies' various campaigns against war is placed prominently on the black-and-white tile floor. "My father and mother thought all sides could enter into a dialogue and come together, but it didn't develop that way," says Hélène with a shrug. Here with its swirling Victorian legs is the massive piano on which Eve practiced to Marie's annoyance. One can almost hear the plunk of tennis balls from the court adjoining the house and the laughter as Frédéric jumps over the net at the end of a set.

Behind the house, the flowers and the trees Marie and Irène so lovingly planted have matured. Close to the house, under a large tree, Marie once sat—one arm thrust around toddler

Frédéric Joliot-Curie and Irène Joliot-Curie in Sceaux.

Eve, the other around Irène's waist. I think this is the photograph that, as a young girl, I tacked up on my bulletin board. "Haven't we all had wonderful lives?" says Hélène, as if reading my thoughts. "My mother told me she had the most interesting life one could imagine. It was my grandmother who made her feel that way."

Yes, Marie Curie's scientific accomplishments and life were remarkable. She was unfettered by the skepticism of fellow scientists, or the fact that she lived in a world where men made the rules. She had a tragic and glorious life. In her own words,

> I am among those who think that science has great beauty. A scientist in his laboratory is not only a technician, he is also a child placed before natural phenomena, which impress him like a fairy tale. We should not allow it to be believed that all scientific progresses can be reduced to mechanism. . . . Neither do I believe that the spirit of adventure runs any risk of disappearing in our world. If I see anything vital around me, it is precisely that spirit of adventure, which seems indestructible.

Acknowledgments

In the mid-1990s Eve Curie Labouisse, Hélène Langevin-Joliot, and Pierre Joliot decided to donate the Curie publications, diaries, journals, and workbooks to the Bibliothèque Nationale on the rue Richelieu. The officials in charge readily accepted, but when Hélène casually mentioned that after seventy-five years these papers were still radioactive, "You should have seen the look on their faces. They were used to dealing with rare manuscripts, and all of a sudden here we were!" Library personnel wrapped each item in a plastic cover, a woefully inadequate solution to the problem. Shortly thereafter, Hélène Langevin-Joliot arrived with a team from the University of Paris at Orsay and Geiger counters. As the counters ticka-ticka-ticked away, the team sorted the books and papers from the Curie Institute into three categories, based on their radioactivity. The most radioactive materials were taken to Orsay for decontamination, a process that took two years to complete. Even so, traces of radioactivity still remain. For several years thereafter researchers were asked to sign a medical release before being permitted to see or handle this material (a practice discontinued in 1999).

In any case, the Curie papers are still not easy to access. Letters of introduction needed to be secured. Mine were handwritten, as is the preferred French custom, by such eminent French experts as Princeton University historian and author Robert Darnton, and New York Public Library president and Voltaire scholar, Paul LeClerc, for which I am most appreciative. Also, my thanks to Jean-Pierre Angremy, the then director of the Bibliothèque Nationale.

The first time I tried to access the Curie papers, I was turned away by the curator, who told me that, though the papers were no longer sealed, I needed a letter from a direct Curie descendant to read and freely quote from them. The following morning, I visited Hélène Langevin-Joliot for the first time. Our meeting was stimulating and cordial. Finally, I asked her if she would be kind enough to write me a note so that I could quote from the Curie papers. "Ridiculous!" she exclaimed. "You don't need a note, these papers are in the public domain. About once a month I get a call asking about this. It is greatly annoying." She seemed quite agitated. Unsure of the next step, I asked, "Well then, could you write me a note saying that I don't need a note?" We both laughed, and that was the beginning of her help, which has been boundless, as a Curie family historian, scientist, and firsthand observer.

Eve Curie Labouisse also has my gratitude for her insight into Marie Curie. Her exemplary biography, *Madame Curie*, was translated into thirty-two languages and won the National Book Award. It was written sixty-seven years ago. After Eve Curie completed it in 1937, the Curie papers were sealed. I believe that most of the material that was written during the extended sealed period, though voluminous, is nearly useless. Since the release of the Curie papers, I believe that the biography *Marie Curie: A Life,* by Susan Quinn, properly utilizes this hidden treasure. Marie's daughter and granddaughter helped me to understand the psychological and scientific aspects of my subject. So did the book by the Curie Institute's Soraya Boudia, *Marie Curie et son laboratoire,* which is unique in exploring the later metrology at which Madame Curie excelled as well as her accomplishments after Pierre's death. Rosalynd Pflaum's two books on the Curie family were also helpful.

I am indebted to Edwin Barber of W. W. Norton & Company; my editors James Atlas and Jesse Cohen; Linda Amster, who repeatedly read the manuscript; Susan Middleton, a superb copy editor; Jeremy Steinke for his invaluable assistance; and most particularly Jason Epstein. Also, my appreciation goes to Suzanne Fedunok, director of the Coles Science Center at New York University's Bobst Library; Mark Piel of the New York Society Library; Michele Sacquin, the curator of the Curie Archive at the Bibliothèque Nationale. Yvonne Baby and Antoine Spire gave me an expert firsthand political assessment of the period following World War II.

I am especially grateful to Dr. Spencer Weart as well as Dr. James A. Carr, Alan Lightman, Dr. Allen Mincer, and Dr. Henry Stroke, who reviewed my manuscript for scientific clarity and accuracy. Dr. Stefan Stein provided an analysis of the roots and development of Marie Curie's depressions. Patricia Osborne, Sébastien Trotignon, Yves-André Istel, and Susan Quinn's biography provided the translations I use from obscure letters, documents, scientific books, articles, and other Curie papers. To all these and more I am deeply grateful.

BARBARA GOLDSMITH, 2004

Notes

Quotations in this book rely heavily on primary sources in the Curie Archive at the Bibliothèque Nationale on the rue Richelieu in Paris, and the Curie and Joliot-Curie Archives at the Institut Curie, as well as on Eve Curie's *Madame Curie*, which was recorded from firsthand observation. In several cases I cite a primary source followed by the secondary source. All quotes not given specific attribution below are taken from the diaries, letters, workbooks, autobiography, and similar papers in the Curie Archive at the Bibliothèque Nationale.

Abbreviations in the Notes are as follows:

BN—Curie Archive, Bibliothèque Nationale, rue Richelieu, Paris.

CI—Curie and Joliot-Curie Archives, Institut Curie, Paris.

MC—Marie Curie.

MC by EC—Eve Curie, *Madame Curie: A Biography by Eve Curie*, trans. Vincent Sheean (Garden City, NJ: Doubleday, Doran, 1937).

PC by MC—Marie Curie, *Pierre Curie*, with an Introduction by Mrs. William Brown Meloney and Autobiographical Notes by Marie Curie (New York: Macmillan, 1923).

Introduction

p. 13   *Suffering from cancer . . . :* Marie Curie was the first woman to be interred in the Panthéon for her own accomplishments; the only other is Sophie Bertholet, who was interred with her husband, the chemist Marcellin Bertholet.

p. 14   *This transfer of Pierre . . .* : Pierre Radvanyi, "Les Curies, Deux couples radioactifs," *Pour la Science: Édition française de Scientific American,* November 2001–February 2002, p. 4.

p. 15   *"TO GREAT MEN FROM . . ."*: "Aux Grands Hommes La Patrie Reconnaisante" in the original.

p. 15   *"noise and ceremonies . . ."*: Soraya Boudia, *Marie Curie et son laboratoire* (Paris: Éditions des archives contemporaines, 2001), p. 65. Direct translation from the French by Patricia Osbourne.

Chapter 1: Early Influences

p. 20   *"several shelves laden . . ."*: *MC* by EC.

p. 20   *Marya Salomee Sklodowska . . .* : Polish nicknames are common. At various times Manya Salomee was called Manya, Maria, Maniusia, Manyusya, and Marya. For the purpose of continuity, I will use Manya, the name her family most frequently used.

p. 21   *"my father had no . . ."*: MC autobiography written in English in her own handwriting, BN.

p. 24   *It was in 1867 . . .* : In Susan Quinn's biography, *Marie Curie: A Life* (New York: Simon & Schuster, 1995), Novolipki is spelled Nowolipki, but throughout the book I have adopted the spellings used in *MC* by EC.

p. 30   *"to scratch like a cat . . ."*: MC autobiography, BN; quoted in *PC* by MC.

p. 30   *"I feel you look down on me"*: *MC* by EC, p. 34.

p. 31   *"many equally childish things"*: Manya to Jozef Sklodowski, BN.

p. 31   *"Sometimes I laugh all by myself . . ."* : Manya to Katzia Przyborovska, BN; quoted in *MC* by EC, p. 40.

Chapter 2: "I Came Through It All Honestly"

p. 35   *"I still believe . . ."*: *PC* by MC, p. 168.

p. 35   *"the Flying University"*: After checking with several scholars on the subject, I find Quinn's use of "the Flying University" (in *Marie Curie*) more apt than Eve Curie's "Floating University" (in *MC* by EC).

p. 36   *"was exactly as enthusiastic . . ."*: *MC* by EC, p. 60.

p. 36   *"One must not enter . . ."*: *MC* by EC, p. 67.

p. 37   "*They were already speaking . . .*": Marie Curie to Bronya, BN.

p. 38   "*stupid incessant parties here . . .*": Ibid.

p. 38   "*There is no other . . .*": The discussion about the incongruence between a Victorian governess' status in family and society comes from the essay "Suffer and Be Still" by Lady Elizabeth Eastlake, 1848; quoted on p. 11 of M. Jeanne Peterson, "The Victorian Governess: Status Incongruence in Family and Society," in *Suffer and Be Still: Women in the Victorian Age*, ed. Martha Vicinus, 3–19 (Bloomington: Indiana University Press, 1972).

p. 38   "*Daniel's Physics, of which . . .*": *MC* by EC, p. 72; quoted in Quinn, *Marie Curie*, p. 70.

p. 38   "*acquired the habit of . . .*": Curie Archive, BN.

p. 39   "*Some people pretend that . . .*": *MC* by EC, p. 72.

p. 40   "*If you only knew . . .*": Ibid., p. 78.

p. 40   "*I can imagine how . . .*": Ibid., p. 77; quoted in Quinn, *Marie Curie*, p. 75.

p. 40   "*I have fallen into black melancholy*": MC letters, BN.

p. 41   "*For me I am very . . .*": Ibid.

p. 43   *in the Tatra Mountains*: The Tatras form the central and most beautiful section of the fifteen-kilometer Carpathian mountain range.

p. 43   "*How funny it would . . .*": MC letters, BN.

p. 43   "*If you can't see . . .*": Ibid.

p. 43   "*Everybody says I have . . .*": Marie Curie to her cousin, Henrietta Michalovska, BN.

Chapter 3: Paris

p. 45   "*We should be interested . . .*": *PC* by MC. MC autobiography and MC papers, BN.

p. 46   "*very precious sense of . . .*": Pierre Curie to Georges Gouy, BN.

p. 48   "*one of the best . . .*": *PC* by MC, p. 170.

p. 48   *The most famous of . . .* : Jules-Henri Poincaré also was the first scientist to suggest a theory of chaos, but lacking today's technology he was unable to prove it.

p. 49   "*All my mind was . . .*": MC autobiography and MC letters, BN.

p. 50   *A French critic wrote . . .* : Gustave Planche.

p. 51   "*The nearer the examinations . . .*": MC autobiography and MC letters, BN.

Chapter 4: Pierre

p. 54   "*Women much more than . . .*": Pierre Curie's journal; quoted in
          PC by MC, p. 77.

p. 55   "*While stretching we observe . . .*": MC workbook, BN. Radvanyi,
          "Les Curies."

p. 56   "*woman of genius,*" "*his nature and his soul . . .*": Pierre Curie cor-
          respondence, letters to MC, BN; both quotes in *PC* by MC, p. 14.

p. 60   "*I wonder if when . . .*": CI; the italics are mine.

Chapter 5: Remarkable Accidents

p. 62   *This was attached to . . .* : The names *cathode* and *anode* are often
          credited to Michael Faraday, but are more frequently credited to
          Eugen Goldstein.

p. 62   *That faraway screen now . . . .* : Fluorescent substances glow while
          being charged with high-energy particles from natural or manu-
          factured sources and cease to glow when the stimulus is removed.
          Luminous substances glow from stimuli other than being heated.
          Phosphorescent substances continue to glow after the stimulus
          ceases.

p. 63   *The bones of Bertha's . . . .* : Most texts say Bertha was wearing a
          wedding ring since it was on the third finger of her left hand,
          but it appears above the knuckle, and at Bobst Library at New
          York University in a similar shadowgram the ring is on the pinkie
          finger. Photographic archives, New York University.

p. 67   "*expecting to find the . . .*": Abraham Pais, *Inward Bound* (New
          York: Oxford University Press, 1986). Also in Alfred Romer, ed.,
          The Discovery of Radioactivity and Transmutation (New York:
          Dover Publications, 1964).

p. 67   "*a lifespan which could be . . .*": A. S. Eve, *Rutherford: Being the
          Life and Letters of the Rt. Hon. Lord Rutherford, O.M.* (New York:
          Macmillan, 1939), p. 78.

p. 67   "*squeezed dry*": Boudia, *Marie Curie et son laboratoire*, p. 33.

Chapter 6: "The Question Was Entirely New"

p. 69   "*the precision of the procedural . . .*": Irène Joliot-Curie commen-
          tary on the Curie notebooks; quoted in *PC* by MC.

p. 75   "*I examined all the . . .*": *PC* by MC, p. 96.

p. 75   *A periodic table devised . . .* : Rosalynd Pflaum, *Grand Obsession:
          Madame Curie and Her World* (New York: Doubleday, 1989), p.

67. Several scientists felt that Mendeléev's table had been inspired by a popular game of the day called "Patience." The form of his chart mimicked this game by writing the names of every known natural element on a blank card then placing them in a horizontal row of eight. With the ninth card he started a new row, thus creating both vertical and horizontal columns.

p. 75 *Mendeléev subscribed to the . . .* : Atomic number relates to the number of protons in the nucleus of the atom. Atomic mass represents the combined number of protons and neutrons.

p. 76 *Henry G. J. Moseley . . .* : As we now know, the atomic weight (more accurately called atomic mass) of an atom represents the combined number of protons and neutrons, while it is the number of electrons—equal to the number of protons—that determines its chemistry.

p. 76 *"It was necessary at . . ."*: Paper written by Marie and Pierre Curie and read by Gabriel Lippmann at the French Academy of Sciences. At first *radio-activity* hyphenated; later it was contracted to *radioactivity*.

p. 77 *"The properties of a metal . . ."*: Dr. Spencer Weart, director of the Center for the History of Physics at the American Institute of Physics, College Park, MD.

p. 78 *"I was brought up . . ."*: Eve, *Rutherford*, p. 384.

p. 78 *The word* atom *itself . . .* : Abraham Pais, *Inward Bound* (New York: Oxford University Press, 1986).

p. 79 *"Though ancient systems may . . ."*: Lawrence Badash, "Radioactivity Before the Curies," chap. in *Radioactivity in America: Growth and Decay of a Science* (Baltimore: Johns Hopkins University Press, 1979).

Chapter 7: "The Best Sprinters"

p. 80 *"I've dug my last potato"*: Eve, *Rutherford*, p. 42.

p. 81 *When J. J. Thomson . . .* : J. J. Thomson should not be confused with Lord Kelvin, William Thomson.

p. 82 *"At first there were . . ."*: J. J. Thomson papers, 1899, Rayleigh Library, University of Cambridge, England.

p. 83 *Both gamma rays and . . .* : There has long been debate as to whether it was Rutherford or Villard who discovered gamma rays. In 1904, Rutherford wrote of his discovery in his book *Radioactivity* (Cambridge: Cambridge University Press, 1904). In 1908, however, when Villard won the Nobel Prize, he claimed both the discovery and the naming of gamma rays.

p. 83   "*I have to publish . . .*": Eve, *Rutherford*, p. 109.

p. 83   "*I had a passionate desire . . .*": MC autobiography, BN.

p. 84   *Pierre . . . "abandoned his work . . .*": PC by MC.

p. 84   "*by the ordinary means . . .*": Radvanyi, "Les Curies," p. 18.

p. 84   "*We soon recognized that . . .*" PC by MC, p. 98.

p. 86   "*obtained a substance 400 . . .*": "Note de M. P. Curie et de Mme. P. Curie sur une nouvelle substance fortement radio-active, contenue dans la pechblende," *Les Comptes rendus de l'Académie des sciences* 127 (1898): 175–178; direct translation by Sébastien Trotignon. Also in Quinn, *Marie Curie*, p. 151.

p. 87   "*If the existence of . . .*": Ibid.

p. 88   *This time he found . . .* : Quinn, *Marie Curie*, p. 152.

p. 88   "*colossal achievement*": Marie Mattingly Meloney, "The Greatest Woman in the World: Marie Curie," *Delineator* (New York), April 1, 1921.

p. 88   *But, in fact, her . . .* : Rutherford and Soddy had proved that radioactivity emanated from inside the atom and that radioactive elements were constantly undergoing a decay process in which they were transmuted into other elements finally to become lead. Quinn, *Marie Curie*, p. 152.

Chapter 8: "A Beautiful Color"

pp. 89–96 Quotes on these pages come from the MC autobiography, BN; quoted in *PC* by MC.

p. 93   *André Debierne was put . . .* : In 1898, he discovered an intensely radioactive element on his own. Actinium, as he called it (coming from the Greek word *aktinos* meaning "ray"), was of little use being the scarcest of all the elements.

p. 97   "*You hardly eat at . . .*": Georges Sagnac to Pierre Curie, Curie letters, BN.

p. 99   *Using Einstein's equation . . .* : David Bodanis, $E = mc^2$: *A Biography of the World's Most Famous Equation* (New York: Berkley Books, 2000), p. 77.

p. 100   "*My father, who in . . .*": CI; quoted in Boudia, *Marie Curie et son laboratoire*.

p. 100   "*You are now in . . .*": Wladyslaw Sklodowski to Manya, CI and BN.

Chapter 9: "What Is the Source of the Energy?"

pp. 102–6 Quotes on these pages not specifically cited below come from the following sources: Contemporaneous papers. Eve, *Rutherford*. David Wilson, *Rutherford: Simple Genius* (Cambridge, MA: MIT Press, 1983). Ernest Rutherford, *The Collected Papers*, vol. 1 (Allen & Unwin, 1962).

p. 102   *In their paper . . . :* This was written with André Debierne.

p. 105   *"This is transmutation . . .":* The gas turned out to be not argon but another inert gas, thoron (a product of the disintegration of thorium), which dissipated in minutes.

p. 105   *"For Mike's Sake, Soddy . . .":* Frederick J. Soddy, "Radioactivity and Atomic Theory," *Journal of Chemical Society* (1902): 840.

p. 106   *"radioactive change must . . .":* Rutherford, *Collected Papers*, p. 43; quoted in Richard Rhodes, *The Making of the Atomic Bomb* (New York: A Touchstone Book / Simon & Schuster, 1986).

Chapter 10: "I Will Make Him an Help Meet for Him"

p. 107   *"the influence of magnetism . . .":* From Lorentz and Zeeman's Nobel Prize presentation speech.

p. 108   *"worked together and separately . . ." :* Letter nominating Pierre Curie and Henri Becquerel for the Nobel Prize.

p. 108   *The most shocking of . . . :* Lippmann was the same man who had presented to the French Academy of Sciences Marie's first paper on radioactivity, who had arranged for her to receive 600 francs for her study of steel, and who had been one of the professors on the defending committee for her doctoral thesis.

p. 109   *"opening up a new . . .":* Report of Knut Ångström to the Committee, Royal Swedish Academy of Sciences, 1903.

p. 109   *"have sometimes been overtaken . . .":* Ibid., p. 189.

p. 111   *"The great success of . . .":* The last sentence in this quote is from Genesis 2:18.

Chapter 11: "The Disaster of Our Lives"

p. 115   *"From early youth Pierre . . .":* PC by MC.

p. 124   *"What sort of new . . .":* MC by EC, p. 359.

p. 124   *"It goes without saying . . .":* Giesel took his radium sample down a deep mine shaft to see if it altered in any respect. It did not. After that, Giesel and Pierre Curie joined forces to investigate the radioactivity of various water sources.

Chapter 12: "We Were Happy"

p. 128   *"some fool in a laboratory . . .":* Wilson, *Rutherford.*

p. 129   *"It is probable that . . .":* Soddy, ""Radioactivity and Atomic Theory."

p. 129   *"Radium and radioactivity have . . .":* Quoted in Oliver Sacks, *Uncle Tungsten: Memories of a Chemical Boyhood* (New York: Alfred A. Knopf, 2001), p. 291.

p. 131   *"It is necessary to . . .":* MC diary, BN; translation provided by Sébastien Trotignon and checked against Quinn, *Marie Curie,* p. 243.

p. 132   *"nothing was going to . . .":* Ibid.

p. 132   *"you left for the laboratory . . .":* Ibid.

p. 133   *"I don't know . . . Don't . . .":* Ibid.

Chapter 13: Metamorphosis

p. 135   *"Pierre is dead . . .":* MC by EC, p. 246.

p. 135   *"elbows on her knees . . .":* Ibid., p. 247.

p. 135   *"It is commonplace to . . .":* Ibid.

p. 136   *One page has been . . . :* We have no idea if these pages were destroyed by Marie or by others.

p. 137   *"intellect . . . cold and . . .":* William Crookes, *Researches in the Phenomena of Spiritualism* (London: J. Burns, 1874).

p. 138   *"thought it possible to . . .":* Anna Hurwic, *Pierre Curie* (Paris: Flammarion, 1998), p. 247.

p. 138   *"I hope we are . . .":* PC letter to Georges Gouy, July 24, 1905, BN.

p. 138   *"There is here in . . .":* PC to Georges Gouy, BN, CI, and the School of Industrial Physics and Chemistry (EPCI).

p. 139   *"I put my head . . .":* MC diary, BN.

p. 140   *"I sometimes have the . . .":* Ibid.

p. 140   *"When you left, the . . .":* Ibid.

p. 140   *"I tried to make . . .":* This was Sunday, April 22, three days after the accident, and her first day back in her lab. MC diary, BN; direct translation by Sébastien Trotignon.

p. 141   *walk as if . . .":* MC diary, p. 18; quoted in *MC* by EC, p. 252ff.

p. 142   *"I want to tell you . . .":* MC diary, BN, p. 25.

p. 142   *"the house, the children . . .":* Ibid., p. 26.

p. 142   *"They have offered that . . .":* MC journal, BN.

p. 143   *"When we examine our . . .":* Radvanyi, "Les Curies," p. 35.

p. 144   *"Yesterday I gave the . . .":* MC diary, BN, p. 28.

Chapter 14: "My Children . . . Cannot Awaken Life in Me"

p. 145   *"I'd never written a . . ."*: From the author's discussion with Eve Curie Labouisse.

p. 145   *"I called the book Madame Curie . . ."*: Ibid.

p. 146   *"Eve was still too . . ."*: MC by EC, p. 266.

p. 146   *"cold as a herring"*: Bodanis, $E = mc^2$.

pp. 147–53 Quotes on these pages are derived from primary sources at the BN, correspondence between Irène Curie and Eve Curie and their mother as noted in the text, as well as from my discussions with Eve Curie, also from *MC* by EC, and Quinn.

Chapter 15: "The Chemistry of the Invisible"

p. 157   *Her name was Lise . . .* : Ruth Lewin Sime, *Lise Meitner: A Life in Physics* (Berkeley: University of California Press, 1996). Sharon Bertsch McGrayne, *Nobel Prize Women in Science: Their Lives, Struggles, and Momentous Discoveries* (Washington, DC: Joseph Henry Press, 1998).

p. 158   *Within a decade Meitner . . .* : The initial laboratory bore the Kaiser's name but after World War II was renamed the Max Planck Institutes.

p. 159   *"For measurements dealing with . . ."*: Boudia, CI mission statement.

p. 160   *"The Madame is not . . ."*: Wilson, *Rutherford*, pp. 256–257.

p. 161   *"for sentimental reasons . . ."*: Ibid., p. 257.

p. 161   *"handle . . . this prickly woman"*: Ibid., p. 254.

p. 162   *"All the eminent women . . ."*: An interview of Sir William Ramsay, *Daily Mail* (England), 1910.

p. 162   *For her there was only . . .* : Pflaum, *Grand Obsession*, p. 140.

p. 163   *"has never been repeated . . ."*: Marie Curie paper on radioactive metal, CI.

p. 163   *"The poor woman has . . ."*: Wilson, *Rutherford*, p. 257.

p. 163   *"It is astonishing how . . ."*: Ibid., p. 256.

p. 164   *"a passionate researcher . . ."*: Eugénie Cotton, *Les Curies* (Paris: Éditions Seghers, 1963), p. 74.

Chapter 16: Honor and Dishonor

p. 166   *Hertha Ayrton . . .* : Hertha Aytron was the wife of the physicist William Allen Ayrton.

pp. 166–70 All of the correspondence and other material in this section have appeared in contemporaneous books and newspapers in several languages with the exception of English language papers, which provided little or no coverage. All of the quotes not otherwise referenced below appeared in the November 23, 1911, issue of *L'Oeuvre*.

p. 168   *"This illustrious woman had . . .":* Jean Perrin testimonial, July 1914, School of Industrial Physics and Chemistry (EPCI).

p. 168   *"Madame Curie looked very . . .":* Wilson, *Rutherford*, p. 138.

p. 169   *"You and I are . . .":* Marguerite Borel [Camille Marbo, pseud.], *À travers deux siècles: 1883–1967* (Paris, 1968): Borel's memoir, written under a pseudonym.

p. 172   *"producing sufficiently pure samples . . .":* Some scholars say that she did not receive the Nobel telegram until November 5, others say November 7. In any case, as future events show, it is a fair assumption that the Nobel Committee did not know of the scandal when this prize was offered.

p. 175   *In 1911 every day . . . :* Françoise Giroud, *Marie Curie: A Life*, trans. Lydia Davis (New York: Holmes & Meier, 1986), p. 165.

Chapter 17: "She Is Very Obstinate"

pp. 177–82 Quotes on these pages are derived from the Curie Archive, as quoted in Eve, *Rutherford*; Wilson, *Rutherford*; and Rutherford, *Collected Papers*.

p. 187   *"She is very obstinate . . .":* Badash, *Rutherford and Boltwood: Letters on Radioactivity* (New Haven: Yale University Press, 1969), p. 75.

Chapter 18: "All My Strength"

p. 184   *"About the trip to . . .":* Correspondence between Barbara Goldsmith and Dr. Hélène Langevin-Joliot, 2003–2004.

p. 185   *"a military installation with . . .":* Giroud, *Marie Curie*, p. 212.

p. 185   *"Only through peaceful means . . .":* MC to Hertha Ayrton, BN.

p. 185   *"It is hard to think . . .":* Ibid., BN.

p. 186   *"resolved to put all . . .":* MC autobiography, BN. Also *Die Presse*.

p. 187   *"It pains me to . . .":* Irène Curie to MC, BN.

p. 188   *"I spent my birthday . . .":* Irène Curie to MC, BN.

Chapter 19: The Making of a Myth

p. 194   *"I understand that it . . .":* MC to Missy Meloney, BN.

p. 195   *"Eve of the radium eyes . . .":* New York Daily News, May 1921.

p. 196   *"a simple woman, working . . .":* Delineator, April 1921.

p. 196   *"On one morning in . . .":* Delineator, April 1921.

p. 197   *"Its glass roof did . . .":* MC autobiography, BN; quoted in *PC* by
          MC.

p. 198   *"Sometimes I had to . . .":* Ibid., p. 186.

p. 198   *"One year would probably . . .":* Ibid., p. 188.

p. 198   *"I was very thankful . . .":* Ibid., p. 224.

p. 200   *"I love it [radium], but . . .":* Henri Becquerel to MC, BN.

p. 200   *"one will feel absolutely nothing . . .":* MC autobiography, BN;
          quoted in *PC* by MC.

p. 200   *"I am happy with my . . .":* Pierre Curie letter, BN.

p. 202   *In the first textbook . . . :* R. Dessauer and B. Wiesner, eds.,
          Radiotreatment (Berlin: von Vogel and Krienbrink, 1904).

p. 203   *Dr. R. F. Mould described . . . : British Journal of Radiology,* 68
          (1995). The description is from the 1930s.

p. 204   *Today a complicated array . . . :* Megavoltage, computer treatment
          planning, SI (Système Internationale) units, conformal therapy,
          mixed-modality treatments (with photons and electrons, for
          example), and remote afterloading using iridium-192, to name
          a few.

Chapter 20: To Pass the Torch

p. 205   *"I hope . . . our Evette . . .":* MC to Irène Curie, BN.

p. 209   *Frédéric began to study . . . :* The cloud chamber was invented by
          C. T. R. Wilson in 1912.

p. 210   *has produced no such . . . :* Spencer R. Weart, *Scientists in Power*
          (Cambridge, MA: Harvard University Press, 1979), p. 44.

p. 212   *"You must get so . . .": MC* by EC, p. 270.

p. 212   *"her mother's old fashioned . . . methodical . . . long . . .":*
          McGrayne, *Nobel Prize Women in Science,* p. 48.

p. 212   *Frédéric picked up a . . . :* In 1928, Hans Geiger invented this
          detector of radioactive particles.

p. 216   *"In the biological sciences . . .":* From the Nobel Prize lecture by
          Pierre Curie.

p. 218   *"a highly dangerous operation":* Quinn, *Marie Curie,* p. 113.

p. 219   *"our occupational disease":* Pierre Biquard, *Frédéric Joliot-Curie:
          The Man and His Theories,* trans. Geoffrey Strachan (New York:
          Paul S. Eriksson, 1966), p. 162.

Chapter 21: Marie's Legacy

p. 222   *He bombarded uranium with . . . :* The expression "off the chart" originally referred to numbers higher than those Mendeléev had predicted for heavy elements.

p. 224   *"I feel so completely . . .":* Sime, *Lise Meitner*, p. 211.

p. 224   *"Perhaps you can propose . . .":* Otto Hahn to Lise Meitner, Meitner correspondence; quoted in McGrayne, *Nobel Prize Women in Science.*

p. 225   *The immense energy released . . . :* Atomic science is now commonly referred to as nuclear chemistry.

p. 226   *"In discovering artificial radioactivity . . .":* The author's interview with Hélène Langevin-Joliot.

p. 226   *"What idiots we have . . .":* Stefan Rozental, ed., *Niels Bohr: His Life and Work as Seen by His Friends* (Amsterdam: North-Holland Publishing, 1967), p. 145.

p. 229   *"We don't realize that . . .":* A speech by Irène Joliot-Curie, 1945, CI.

p. 230   *"nuclear energy has only . . .":* The author's interview with Hélène Langevin-Joliot.

p. 230   *"In those days school . . .":* Ibid.

p. 231   *In 1954, Hélène helped . . . :* The scintillation spectrometer was designed to analyze photons of intense electron radiation. These photons induced fission, notably in uranium-238 (photo fission).

p. 231   *Two years later she . . . :* The title of her thesis was "Internal Bremmstrahlung and the Auto-ionization Phenomenon." Bremmstrahlung (literally "braking radiation") consists of X-rays or gamma rays (depending on the amount of energy) emitted by an electron as it comes to a stop due to a sudden collision. In this case the collision takes place inside a radioactive atom, whereas in an X-ray tube it takes place when electrons hit the "target" inside the tube. For her thesis Langevin-Joliot also studied autoionization, which led to a modification of the existing theory. Hélène Langevin-Joliot thesis and summary, Science Archives, Coles Science Center, Bobst Library, New York University.

p. 233   *"I am among those . . .":* MC archive, BN; quoted in *MC* by EC, p. 341.

Selected Bibliography

Except for several articles listed below, I do not list or cite the hundreds of other scientific articles, theses, papers, and other works that were used in addition to these sources.

Books and Articles

Badash, Lawrence. *Radioactivity in America: Growth and Decay of a Science.* Baltimore: Johns Hopkins University Press, 1979.

———, ed. *Rutherford and Boltwood: Letters on Radioactivity.* New Haven: Yale University Press, 1969.

Barnes-Svarney, Patricia, ed. dir. *The New York Public Library Science Desk Reference.* The New York Public Library Series. New York: A Stonesong Press Book / Macmillan, 1995.

Bigland, Eileen. *Madame Curie.* New York: S. G. Phillips, 1957.

Biquard, Pierre. *Frédéric Joliot-Curie: The Man and His Theories.* Translated by Geoffrey Strachan. New York: Paul S. Eriksson, 1966.

Birch, Beverley. *Marie Curie: Courageous Pioneer in the Study of Radioactivity.* Woodbridge, CT: Blackbirch Press, 2000.

Birch, Beverley, and Christian Birmingham. *Marie Curie's Search for Radium*. Hauppauge, NY: Barron's Educational Series, 1995.

Bodanis, David. *E = mc²: A Biography of the World's Most Famous Equation*. New York: Berkley Books, 2000.

Borel, Marguerite [Camille Marbo, pseud]. *À travers deux siècles: 1883–1967*. Paris, 1968. (Borel's memoir, written under a pseudonym.)

Boudia, Soraya. *Marie Curie et son laboratoire*. Paris: Éditions des archives contemporaines, 2001. (Direct translation from the French by Patricia Osbourne.)

———. "The Curie Laboratory: Radioactivity and Metrology." *History and Technology* 13 (1997): 249–265.

Bradshaw, Louis. "Understanding Piezoelectric Quartz Crystals." *RF Design*. August 2000. http://home.hetnet.nl/~pasopd/pdfs/piezoelectric.pdf (www.rfdesign.com).

Bragg, Melvyn, with Ruth Gardiner. *On Giants Shoulders: Great Scientists and Their Discoveries—From Archimedes to DNA*. New York: John Wiley & Sons, 1998.

Campbell, John. *Rutherford Scientist Supreme*. Christchurch, N.Z.: AAS Publications, 1999.

Cotton, Eugénie. *Les Curies*. Paris: Éditions Seghers, 1963.

Crawford, Elizabeth. *The Beginnings of the Nobel Institution: The Science Prizes, 1901–1915*. Cambridge: Cambridge University Press, 1984.

Crookes, William. *Researches in the Phenomena of Spiritualism*. London: J. Burns, 1874.

Curie, Eve. *Madame Curie: A Biography by Eve Curie*. Translated by Vincent Sheean. Garden City, NJ: Doubleday, Doran, 1937.

Curie, Marie. *La Radiologie et la guerre* [Radiology and the War]. Paris: Librairie Félix Alcan, 1921.

———. *L'Isotopie et les éléments isotopes*. Paris: Librairie Scientifique Albert Blanchard, 1924.

———. *Oeuvres de Pierre Curie* [Works of Pierre Curie]. Paris: Gauthier-Villars, 1908.

———. *Pierre Curie*. With an Introduction by Mrs. William Brown Meloney, and Autobiographical Notes by Marie Curie. New York: Macmillan, 1923.

———. "Radioactive Substances." New York: Philosophical Library, 1961.

(A translation from the French of the classical thesis presented to the Faculty of Sciences in Paris.)

——. *Traité de radioactivité* [Treatise on Radioactivity]. 2 vols. Paris: Gauthier-Villars, 1910.

——. *Oeuvres de Pierre Curie* [Works of Pierre Curie]. Paris: Gauthier-Villars, 1908.

Dessauer, R., and B. Wiesner, eds. *Radiotreatment*. Berlin: von Vogel and Krienbrink, 1904.

Dry, Sarah. *Curie*. With an essay by Sabine Seifert. London: Haus Publishing, 2003.

DuBois, Ellen Carol, ed. *Woman Suffrage and Women's Rights*. New York: New York University Press, 1998.

Edwards, Stewart, ed. *The Communards of Paris, 1871*. Ithaca, New York: Cornell University Press, 1985.

Eisenhart, Margaret A., and Elizabeth Finkel. *Women's Science: Learning and Succeeding from the Margins*. Chicago: University of Chicago Press, 1998.

Emsley, John. *Nature's Building Blocks: An A–Z Guide to the Elements*. Oxford: Oxford University Press, 2001.

Eve, A. S. *Rutherford: Being the Life and Letters of the Rt. Hon. Lord Rutherford, O.M.* New York: Macmillan, 1939.

Fermi, Laura. *Atoms in the Family: My Life with Enrico Fermi*. Chicago: University of Chicago Press, 1954.

Fullick, Ann. *Marie Curie*. Chicago: Heinemann Library, 2001.

Giroud, Françoise. *Madame Curie: A Life*. Translated by Lydia Davis. New York: Holmes & Meier, 1986.

Greene, Carol. *Marie Curie: Pioneer Physicist*. Chicago: Children's Press, 1984.

Horvitz, Leslie Alan. *The Quotable Scientist: Words of Wisdom from Charles Darwin, Albert Einstein, Richard Feynman, Galileo, Marie Curie, and More*. New York: McGraw-Hill, 2000.

Hurwic, Anna. *Pierre Curie*. Paris: Flammarion, 1998.

Hurwic, Jozef. "Importance de la thèse de doctorat de Marie Sklodowska Curie pour le développement de la science sur la radioactivité." Thesis, Faculty of Sciences, Paris, 1992.

Joliot-Curie, Irène. "Marie Curie, ma mère," *Europe* 108 (1954): 89–121.

Langevin-Joliot, Hélène. "Radium, Marie Curie and Modern Science." *Radiat Res* 150, no. 5 (1998): S3–S8.

Lepscky, Ibi. *Marie Curie.* Translated by Marcel Danesi. Hauppauge, NY: Barron's Educational Series, 1993.

London Times. "The British Association—A Forecast," September 5, 1896. "The British Association," September 24, 1896. "Science in 1896," January 14, 1897. "Science in 1898: Physiology," January 20, 1899. "The British Association—A Forecast," August 29, 1899. "British Association," September 6, 1900. "The Investigation of Cancer," April 21, 1902. "The Mystery of Radium," March 25, 1903. "The Mystery of Radium" (letter to the editor), April 13, 1903. "Cancer and Its Origin," December 10, 1903. "M. Curie on Radium," February 22, 1904. "Cancer Research," February 25, 1904. "The British Association," August 18, 1904. "Sir Oliver Lodge and Huxley" (letters to the editor), October 25, 1904. "The International Congress for Cancer Research," October 1, 1906. "The Treatment of Cancer," December 15, 1906.

Loriot, Noëlle, with the collaboration of Doctor Houdard-Koessler. *Irène Joliot-Curie: un destin au service de la science* [a destiny in the service of science]. Paris: Presses de la Renaissance, 1991. (Direct translation from the French by Patricia Osbourne.)

McGrayne, Sharon Bertsch. *Nobel Prize Women in Science: Their Lives, Struggles, and Momentous Discoveries.* Washington, DC: Joseph Henry Press, 1998.

McKown, Robin. *She Lived for Science: Irène Joliot-Curie.* New York: Julian Messner, 1961.

Pais, Abraham. "The Discovery of the Electron." *Beam Line*, spring 1997.

———. *Inward Bound.* New York: Oxford University Press, 1986.

Parker, Steve. *Marie Curie and Radium.* Science Discoveries. New York: HarperTrophy, 1992; reprint edition.

Pasachoff, Naomi E. *Marie Curie and the Science of Radioactivity.* Oxford Portraits in Science. New York: Oxford University Press, 1996.

Perrin, Jean. "Madame Curie et la découverte du Radium." *Vient de paraitre* (monthly bibliographical bulletin), February 1924.

Peterson, M. Jeanne, "The Victorian Governess: Status Incongruence in Family and Society." In *Suffer and Be Still: Women in the Victorian Age.*

Edited by Martha Vicinus, 3–19. Bloomington: Indiana University Press, 1972.

Pflaum, Rosalynd. *Grand Obsession: Madame Curie and Her World*. New York: Doubleday, 1989.

———. *Marie Curie and her Daughter Irène*. Minneapolis: Lerner Publications, 1993.

Poynter, Margaret. *Marie Curie: Discoverer of Radium*. Berkeley Heights, NJ: Enslow Publishers, 1994.

Quere, d'Yves. *Leçons de Marie Curie: Recueillies par Isabelle Chavannes en 1907*. Paris: EDP Sciences, 2003.

Quinn, Susan. *Marie Curie: A Life*. New York: Simon & Schuster, 1995

Radvanyi, Pierre. "Les Curie, Deux couples radioactifs." *Pour la Science: Édition française de Scientific American*, November 2001-February 2002.

———, and Monique Bordry. *La Radioactivité artificielle et son histoire*. Paris: Seuil / CNRS, 1984.

Reid, Robert. *Marie Curie*. New York: Saturday Review Press / E. P. Dutton, 1974.

Rhodes, Richard. *The Making of the Atomic Bomb*. New York: A Touchstone Book / Simon & Schuster, 1986.

Romer, Alfred, ed. *The Discovery of Radioactivity and Transmutation*. New York: Dover Publications, 1964.

———. *The Restless Atom*. Garden City: Anchor Books, 1960.

Rossiter, Margaret W. *Women Scientists in America: Struggles and Strategies to 1940*. Baltimore: Johns Hopkins University Press, 1982.

Rozental, Stefan, ed. *Neils Bohr: His Life and Work as Seen by His Friends*. Amsterdam: North-Holland Publishing, 1967.

Rutherford, Ernest. *The Collected Papers*. Volume 1. Allen & Unwin, 1962.

———. *Radioactivity*. Cambridge: Cambridge University Press, 1904.

———, and Frederick Soddy. "The Cause and Nature of Radioactivity." *Philosophical Magazine* 4 (1902): 370–396.

Sacks, Oliver. *Uncle Tungsten: Memories of a Chemical Boyhood*. New York: Alfred A. Knopf, 2001.

Sadoul, Georges. *Mystère et Puissance de l'Atome*. Paris, France: Éditions Hier et Aujourd'hui, 1947.

Senior, John E. *Marie and Pierre Curie*. Stroud, England: Sutton Publishing, 1998.

Seymour, Elaine, and Nancy M. Hewitt. *Talking About Leaving: Why Undergraduates Leave the Sciences.* Boulder, CO: Westview Press, 1997.

Sime, Ruth Lewin. *Lise Meitner: A Life in Physics.* Berkeley: University of California Press, 1996.

Soddy, Frederick J. "Radioactivity and Atomic Theory." *Journal of Chemical Society* (1902).

Steinke, Ann E. *Marie Curie and the Discovery of Radium.* Hauppauge, NY: Barron's Educational Series, 1987.

Szalay, Hélèna Sklodowska. *Ze Wspomnien o Marii Sklodowskiej-Curie* [in Polish]. Translated by Alexandra Gordinier and Anna Sobczynski. Nasza Ksiagarnia, 1958. (Hélèna Sklodowska's memoir.)

Vicinus, Martha, ed. *Suffer and Be Still: Women in the Victorian Age.* Bloomington: Indiana University Press, 1972.

Weart, Spencer R. *Scientists in Power.* Cambridge, MA: Harvard University Press, 1979.

Wilson, David. *Rutherford: Simple Genius.* Cambridge, MA: MIT Press, 1983.

Woznicki, Robert. *Madame Curie: Daughter of Poland.* Miami, FL: American Institute of Polish Culture, 1983.

Zak, Sonia. *Frédéric et Irène Joliot-Curie.* France: Éditions Causette, 2000.

Ziegler, Gillette, ed. *Choix de Lettres de Marie Curie et Irène Joliot-Curie* [Selected Letters . . .]. Paris: Les Éditeurs Français Réunis, 1974.

Major Archival Sources

Bibliothèque Nationale, rue Richelieu, Paris. Départment des Manuscrits.

Institut Curie, Paris. Personal correspondence between Irène and Marie Curie, Irène and Frédéric Joliot-Curie, Paul Langevin and Frédéric Joliot-Curie, Paul Langevin and Irène Joliot-Curie, Missy Meloney and Marie Curie.

New York University Science Library. Scientific archives, articles, and papers.

Journals, Newspapers, Magazines

Acta Mathematica, Sweden

British Journal of Radiology, England
Daily Mail, England
Daily News (The Sunday Magazine), United States
Delineator, United States
Die Presse, Austria
Excelsior, France
Illustration, France
Je sais tout, France
L'Action française
Le Correspondant, France
Le Figaro, France
Le Paris-Journal, France
Le Radium, France
Les Comptes rendus de l'Académie des sciences, France
Le Temps, France
L'Humanité, France
L'Intransigeant, France
L'Oeuvre, France
London Times, England
Paris-Presse, France
Philosophical Magazine, France

Photo Credits

Except where otherwise noted, all illustrations are courtesy of the Curie and Joliot-Curie Association/Curie and Joliot-Curie Fund.

Page 19: Metro-Goldwyn-Mayer publicity still, courtesy of Photofest.

Page 64: National Library of Medicine.

Page 110: *Die Berühmten Erfinder, Physiker, und Ingenieure*, Aulis Verlag Deubner, Köln, Germany.

Page 139: Bibliotheque Nationale, Paris/Archives Charmet/Bridgeman Art Library.

Page 173: ACRPP.